Home Land

Home Land

Ranching and a West That Works

Edited by Laura Pritchett, Richard L. Knight, and Jeff Lee

Introduction by Teresa Jordan

ROCKY MOUNTAIN LAND LIBRARY

Johnson Books, Boulder

Published by Johnson Books, a Big Earth Publishing company,
3005 Center Green Drive, Suite 220, Boulder, Colorado 80301.
E-mail: books@bigearthpublishing.com
www.bigearthpublishing.com
1-800-258-5830

Cover painting by Teresa Jordan
Text design and composition by Michael Cutter

9   8   7   6   5   4   3   2   1

Library of Congress Cataloging-in-Publication Data
Home land: ranching and a West that works / edited by Laura Pritchett, Richard L. Knight, and Jeff Lee.
     p.cm.
ISBN 155566-400-8
1. Ranch life—West(U.S.)--Anecdotes. 2. Ranchers--West (U.S.)--Anecdotes. 3. Conservation of natural resources--West (U.S.) I. Pritchett, Laura, 1971- II. Knight, Richard L. III. Lee, Jeff. IV. Title: Ranching and a West that works.
     F596.H6837 2006
     978--dc22

                          2006034518

Printed in the United States of America

*To those who practice Husbandry and Stewardship
for the good of people, land, and animals*

Contents

Preface

The most tragic conflict in the history of conservation is that between conservationists and farmers and ranchers. It is tragic because it is unnecessary. There is no irresolvable conflict here, but the conflict that exists can be resolved only on the basis of a common understanding of good practice. We need to foster and study working models: farms and ranches that are knowledgeably striving to bring economic practice into line with ecological reality, and local food economies in which consumers conscientiously support the best land stewardship.

—Wendell Berry, *The Whole Horse*

The premise of our book is that conservation is compatible with ranching and farming. For evidence of this, go out and look at the land. In watersheds, basins, rangelands, and mountains across the West, livestock are being managed differently, in innovative and progressive ways. And the land is responding. Vegetation is more diverse and productive, soils are more stable and there is less bare ground, streams and springs have come back to life, and biodiversity thrives.

A second premise of this book is that conservationists and agriculturalists are compatible creatures: they actually belong together. (Yes, we just said that, and yes, we mean it.) There's plenty of meeting ground, of common ground, for people who care about the same basics: stewardship, land health, open space, communities, cultures, people, earth.

The contributors of this book think so too. These poets, award-winning authors, farmers, ranchers, environmentalists, academics, city folk, and rural dwellers will all tell you that ranching, farming, and conservation are inexorably linked; and so are the people involved. For that matter, everyone who eats is involved too. So are those who like their views. So are the people who give a damn about nearly anything out in the West, because it's all connected, of course.

Specifically, this book offers stories about creating and sustaining land health through the "radical center," a meeting ground where diverse parties can come to discuss their interests instead of arguing their positions. As Joan Chevalier puts it in her essay, we've unfortunately come to a point that

> pressure politics on both sides of the divide has reached its nadir, that those engaged in this conflict ... have become more consumed with winning the argument than with finding solutions, and that the landscape and its people cannot survive much more of this combat. ... The radical center seeks to address the long-term survival of both agriculturalists and landscapes by focusing on solutions that encompass environmental stewardship and economic vitality as mutually reinforcing common and community goods.

With this radical center as a backdrop, we can then reconsider our government's use of the term "homeland," as in "Homeland Security." We titled this book "Home Land" because we wanted people to remember what a homeland *really* is. A secure homeland is not the country with the greatest military. Home, land, and security are about ecologically sustainable food production; about compensating ranchers and farmers fairly for a healthy food product as well as for protecting open space, wildlife habitat, and watersheds through their land stewardship; and about honoring our cultures, our communities, and our land. We hope that you, along with individuals from the broad spectrum of people who care about the West, share our belief that collaboration, not confrontation, is the key to a West that works better.

Rural communities across the West stand at a threshold. For these communities to be healthy and prosperous they must succeed in recovering and maintaining healthy landscapes. But how can rural communities succeed in a global economy when people are becoming increasingly removed from land and when urban Americans seemingly do not care whether food comes from near or from afar?

Urban communities face an equally precarious future, one increasingly divorced from land, the source not only of their pleasures and inspiration, but also the basis of their food, wood products, and energy. In the short term, people may believe that their economies and happiness are independent of the land, but in time this belief becomes a dangerous delusion. To believe that money actually represents food is to lose sight of the timeless truth that healthy food is connected to rich soil, clean water, and people who husband, cultivate, and harvest our most fundamental necessity.

The answer to these dilemmas lies in connecting these necessary products of land to the naturally desirous needs of urban people through healthy food and open space. We all share these desires, and now it is time to find the connections. Our contributors are people who, in Mary Austin's words, "summer and winter with the land." They do not suffer the delusions that Wendell Berry warned us about:

> Most of us cannot imagine the wheat beyond the bread, or the farmer beyond the wheat, or the farm beyond the farmer, or the history beyond the farm. Most people cannot imagine the forest and forest economy that produced their houses and furniture and paper; or the landscapes, the streams and the weather that fill their pitchers and bathtubs and swimming pools with water. Most people appear to assume that when they have paid their money for these things they have entirely met their obligations.

Not only do the authors of this book see beyond the finished food product, they have given us a collection of essays and poems that overlap and extend in their own ways—as they should, for finally, we come to learn a place, and a person, and a culture, by perceiving the *relationships* that connect them all.

To help protect the lands of the West, the authors and editors are donating the royalties from the publication of this book to the Colorado Cattlemen's Agricultural Land Trust, the West's first agricultural land trust, which has partnered with more than 100 ranching families to protect hundreds of thousands of acres. You can find out more at the end of this book.

So thank you for taking the time to pick our book up. We hope it offers you many lasting pleasures. We have no doubt that rebuilding once-vibrant connections across people and land is essential to our health and happiness. After all, what is more fundamental than life and land?

Acknowledgments

First and foremost, we would like to thank our contributors. They have their hands in the soil, their eyes on the hills, and their lives devoted to a better West, a West where people and land get on well together. They remind us all that there is a better way, one that combines the pleasures of connecting people to land.

Laura Pritchett would like to thank Jim and Rose Brinks, her parents, who gave her countless gifts, one of which was moving to a ranch when she was a girl. Also Wendy Stallard Flory, John Calderazzo, Shirley Rose, and Edna Loehman—members of a committee who helped her down this particular path. Thanks also to Jeff Lee and Rick Knight, who became true friends throughout the course of this project. And to James, Jake, and Eliana Pritchett for growing with her under the stars.

Richard L. Knight offers thanks to the good people who inspire him in his work. This includes students at Colorado State University, rural families who have opened their doors to him over the decades, and westerners, who by their actions represent the hope for a better West. Last, to Heather, for love and sweet inspiration.

Jeff Lee is grateful for the hope and inspiration gained from Lynne Sherrod and the Colorado Cattlemen's Agricultural Land Trust—and from the many friends and supporters of the Rocky Mountain Land Library. These are people truly passionate and committed to a West that works.

Thanks to Paul F. Starrs of the University of Nevada, Reno, for the NASA photo of the United States at night in the Introduction.

Thanks also to Mira Perrizo and Johnson Books for their support of this project.

"We stand for what we stand on."

—Wendell Berry

Introduction

Self-Reliance: Cooperation and the Reawakening of the American Spirit

Teresa Jordan

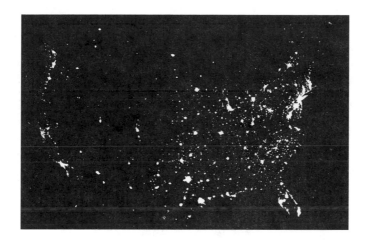

From outer space, the United States west of the hundredth meridian looks almost empty. The lights of the few big cities and the far-flung towns shine out of a blackness so vast it would be easy to think one was looking up at a smattering of stars, instead of down on a major portion of our motherland.

It is precisely because of these vast open spaces that we love the American West. It has been home ground for peoples both ancient and modern, proving ground for generations of adventurers, playground and pressure valve for the ever-increasing stresses of contemporary life. Whether we live in the American West or simply dream of visiting it, we revere it for its fresh air, endless horizons, and wealth of wildlife and natural beauty. Half of the land west of the hundredth meridian is public land: it is quite literally our birthright, a rich and rare gift.

But however serene the West seems from space, it is rife with conflict on the ground. Counties fight states and states fight Washington;

1

newcomers fight longtime residents; environmentalists, ranchers, miners, loggers, and Native tribes fight each other; developers fight anyone who is old and in the way. Like all wars, this one has casualties. A dozen cows are shot in Colorado; a forest ranger's family barely escapes injury when his van explodes in Nevada; two environmentalists are hanged in effigy in Oregon; a logger dies when his saw hits a spike in Washington. Perhaps most damaging of all, in the midst of the fighting, millions of acres of the best watered, most productive, and biologically diverse land are cut up into subdivisions, and those of us who love the West—whether ranchers or environmentalists or the owners of brand new ranchettes—feel our hearts shattered like glass.

We live our lives by stories. The American West has been shaped to a large degree by the stories of conflict and war, of rugged individualists faced off against each other until only one remains. But this war—these many wars—are killing us and they are killing the West we love. There are other stories, stories of cooperation, stories that affirm life and the human spirit. Those are the stories gathered here.

Wallace Stegner once wrote that the guts of any significant fiction—or autobiography—is an anguished question. Several years ago, I published a memoir titled *Riding the White Horse Home* about Iron Mountain, Wyoming, the ranching community in which I was raised. The anguished question that drove the book was "Why, if my people were so strong, were we not resilient?" My family and the other ranchers in Iron Mountain *were* strong. As my grandfather was wont to say, we killed our own snakes and buried our own dead. But we were not resilient. Of the third-, fourth-, and fifth-generation ranches that made up the community of my childhood, only one remains intact. And of course Iron Mountain is not unique. During my lifetime, tens of thousands of ranches comprising millions of acres of land have left family management forever.

I was raised to think that we were the last of the rugged individualists: unflinching, unwhining, unbending. In a word, self-reliant. Over the intervening years, however, I have realized that rugged individualism and self-reliance are not always synonyms but can in fact work against each other. We would rather give help than ask for it, rather be in debt to the bank than indebted to a neighbor. We were proud that we went down fighting, but the bottom line was that we ended up flat on our backs.

Years ago I heard Allan Savory, a native of what is now Zimbabwe and a pioneer in holistic ecosystem management who has worked all over the world, say that he had never experienced individualism as pronounced as he found in the American West. Extreme individualism created a void, he said, and bureaucracy, litigation, and acrimony filled that void. At first I disagreed. My people *hate* bureaucracy or outside interference of any kind. But I had to look no further than my own family to see Savory's thesis played out in practice. My grandfather, Sunny Jordan, had been unwilling to effect the estate planning that would have allowed our ranch to pass from one generation to another upon his death in 1973. Because we hadn't been able to work together as a family, much of his legacy went to lawyers and taxes, and an oil company now owns the land on which he and his father before him had spent their life's blood.

We live our lives by stories. My family lost the family ranch for lots of reasons. Estate taxes were confiscatory; interest rates were high. Under any circumstances, we would have had a hard time. But primary among the causes was that we hadn't been able to plan for the generations. How could this happen? Why had my grandfather made it so hard for my father to carry on? We were not a family at war with one another. We saw ourselves as loving people. But we were shaped by a story. I suspect that most people with a passionate commitment to the West have an underlying story that shapes their passion. This is ours, our creation myth, the story of my great-grandfather's coming to Wyoming.

J.L. Jordan grew up in Maryland. He had wanted to join his brothers fighting the Civil War but he was only fourteen. When his parents denied him permission to go, he ran away, but before he could reach the front, Lee surrendered. Since he had disobeyed his parents, he couldn't return home so he headed out west. He worked his way across the country, building bridges in Illinois and railroads in Nebraska. He ended up in Wyoming and found a job on a ranch. His employer died and J.L.

hadn't been paid for some time. In lieu of wages, he received title to what would become the core of the Jordan ranch. Through the years he bought more cattle and more land, and he never wrote home until he had made himself into a success.

That, at any rate, was the family gospel that I learned at my grandfather Sunny's knee when I asked how the ranch began. But years later as an adult, when I sat down to write the history of the ranch, I remembered a packet of letters that a distant relative back east had sent to my great-aunt Marie (Sunny's sister and J.L.'s daughter) a few years earlier, letters J.L. had written his father back in Maryland. When Marie died, the letters came to me. I had glanced through them enough to recognize familiar details—in one, J.L. mentioned the bridge gang; in another, the railroad crew—but I hadn't really read them. Now I put the letters in chronological order, and the dates leapt out at me. The earliest were written in 1886, over twenty years after Lee surrendered. I consulted the family Bible. J.L. had been born in 1861. He would have been four years old at Appomattox Courthouse.

I started to read the correspondence in earnest. J.L. hadn't run away from home; he left at the age of twenty-five with his family's blessing. He hadn't cut himself off; he chronicled his trip across the country in letters to his father, and he was often homesick. He found his work in Wyoming particularly dismal, with its wind and cold and loneliness. But his boss offered him the chance to buy cattle and his father lent him money to make it possible. Later, when his brothers' sawmill burned down back in Maryland, J.L. lent them money to start up again.

The letters told a story, but not the one I had heard at Sunny's knee. His tale was of a fourteen-year-old orphan who had come west and done it all alone. The real story was of a family that had helped one of its members get established. Once he was successful, he helped them in return.

I cannot describe the revelation that these two stories, laid side by side, provided for me. It was the first time I fully understood how much our stories shape us, how much we are the stories we tell about ourselves. My grandfather had died without making the arrangements that would have allowed the ranch to pass to my father. Years before, my father had moved off the ranch for many years after an argument with his own father, and hadn't moved back until J.L. died. Like many western families—like many American families—mine had a history of fathers fighting sons. And as I thought about the generations of tension that had distanced Sunny from J.L., my father from Sunny, and my brother from the men who preceded him, I thought, How could it have been any different? They had all measured themselves and each other against someone who had never existed.

I'm certain Sunny never consciously changed his father's story. Rather, he absorbed it from the culture at large. Owen Wister's Virginian, the hero of the most popular Western ever written, was a fourteen-year-old orphan, as were many of the heroes of the Horatio Alger stories. These orphans turn up in so many popular novels and have been absorbed into so many family stories that one would think the trails west were blazed by armies of homeless fourteen-year-olds. My family told one story about ourselves: that our progenitor had done it all alone, that each of us should. It severed the bonds between us and put us off the land. Things might have been different if we had told a story somewhat closer to the truth.

The "take no prisoners" approach that limited my family's options is not, of course, limited to family affairs or the ranching industry. It underlies many of the seemingly intractable problems that plague the West. Environmentalists have been at least as rigid as ranchers in the thirty-year war that has locked vast regions of the West in litigation and rancor while millions of acres of open land have fallen to development. Exurban developments themselves are driven by an individualist aesthetic deep in the American soul that craves a piece of land out of sight of the next man's wood smoke. And when an ember escapes the flue, or a cigarette butt or lightning strike sets the prairie ablaze, the failure of a devastated community to minimize their risk in advance often traces back to a distaste for the sort of foresight and cooperation that might have helped them develop a fire prevention plan. Small towns flounder and the surrounding farms and ranches sell on the auction block in part because the food in the local supermarket comes from thousands of miles away rather than from the local economy.

Every day more people sink into despair—and too often into bankruptcy—under the weight of steamrolling forces that seem out of their control: commodity agriculture in the hands of agribusiness syndicates, environmental regulations driven by urban interests, litigation financed by large and distant organizations, bureaucratic gridlock, debt, relentless demographic pressure, wind and drought and fire.

And yet all across the West, pockets of optimism are developing where people with a broad spectrum of interests and beliefs have found ways to work together, creating robust communities ripe with ecological, psychological, spiritual, and economic health. In the process, they have learned to build trust among people who once considered themselves at war. These communities have become self-reliant by transforming the negative aspects of western, rugged individualism into a positive cooperation that affirms the individual. They are forging new stories to live by. I'll tell you one of those stories, for I had the chance to participate in it myself.

In 1993, shortly after my husband and I moved to Nevada, I started working with a ranch restoration project in the Toiyabe Range in central Nevada, close to the old mining town of Austin. The ranchers, Tony and Jerrie Tipton, ran almost entirely on public land. When they took on their allotments in the mid-1980s, the land was in terrible shape. Even though it had been rested for five years when the Tiptons first saw it, there were vast expanses of hard, sunbaked ground and the surviving grasses were largely decadent and dying. Everywhere they looked, they saw erosion and broken-down streambanks. They found evidence of dried-up springs, but few live ones, and those ran thick with mud and silt. Fish, birds, wildlife, and even insects were rare.

Trained in ecosystem-based or holistic ranching, the Tiptons believed they could improve things, but they needed more flexibility than Bureau of Land Management and Forest Service leases allowed. They worked out an agreement with the agencies. If the agencies would loosen restrictions on how many cattle the Tiptons ran, where, and for how long, the Tiptons would give the authority over how that flexibility was used to a collaborative team composed of representatives from the BLM and Forest Service, state agencies such as the Nevada Division of Wildlife and the Nevada Division of Forestry, environmental groups such as the Sierra Club and Trout Unlimited, local ranchers, and anyone else who was interested.

The Toiyabe Watershed and Wildlands Management Team formed and developed goals, which were revised each year, for the improvement of the land base that formed the Tipton Ranch. The team was open to anyone willing to participate, and all members had their interests reflected in the goals. That is to say, if someone wanted improved trout habitat or the reintroduction of blue grouse and the goals didn't already reflect those concerns, they were revised to include them. To move toward these goals, the team developed and constantly revised a management plan that the Tiptons put into action. The team monitored the ranch closely, measuring everything from ground cover to stream flow to wildlife populations to insect activity. The Memorandums of Understanding from the federal agencies that allowed flexibility remained in effect only so long as team members were in agreement—and the monitoring data reflected—that there had been significant improvement. In other words, anyone could derail the process if he or she was unhappy with what was going on and could back up that frustration with evidence from the land.

I first learned about this group when they sponsored a field day. As we toured the ranch, I pored over monitoring data and photographic records, comparing conditions a few years earlier to what I saw before me. I was amazed. Ground cover had increased dramatically—in places by more than 500 percent—and the ratio of perennials to annuals had also increased several-fold. Both of these factors contributed to watershed improvement. Streambeds had raised, banks had stabilized, and in several cases ephemeral streams were flowing year-round while others that had been dry for decades had started to flow or showed signs that they soon would.

With great excitement I joined the team. Here at last seemed to be a solution for the long-smoldering enmities flaring all around me. Ranchers, environmentalists, and government agents were working together, and both the land and the people were the better for it. Looking back at my early involvement with the team, I was downright infatuated. This is it! I'd thought. This is the answer! But then, in the summer of 1994, things started to fall apart.

To move toward its goals of increasing the health and biodiversity of the land, the team had several tools at its disposal, including rest, fire, grazing, animal impact, and technology. Cattle were part of the equation, but they were managed differently than has traditionally been the case, stocking an area more densely but for a much shorter period of time. As the herd moved through, hooves tilled the land, breaking up soil so it could better retain moisture and allow seedlings to establish, while dung and urine provided fertilizer. The cows ate all the grass in their path, but they ate it only once. This "harvest" revitalized the grass, and the cows had moved on before they could eat regrowth, which would have stressed it. This sort of grazing can improve land, increasing the amount of ground cover dramatically and jump-starting succession so that perennial grasses, with their stabilizing and water-storing root systems, succeed over annuals. Although many team members initially opposed cattle, over time they came to see them as an effective tool for ecosystem restoration.

For the process to work, however, the cattle must be carefully controlled. In our biological planning session each year, we plotted where they would be throughout the summer, in what numbers, and for how long. This worked fine in theory, but in the summer of 1994 the cattle failed to read the memo. The Tiptons had introduced new livestock to the ranch that didn't know the lay of the country. No sooner would they move the cattle to a new place than they would head back to the old. Compounding the problem was the fact that, after years of red tape, the Forest Service had finally succeeded in contracting out a five-mile stretch of fence the team had deemed necessary to help with cattle control. Unfortunately, the contract ran out a half-mile from completion, rendering the fence essentially useless. In midsummer, when some team members from the Nevada Division of Wildlife brought visitors to the ranch to show off what they had billed as a great success, they were embarrassed to find cattle on a stream from which they should have been excluded, and the cows had trashed the place. There were other problems of a similar nature: the Forest Service had issued a letter threatening an end to the Memorandum of Understanding; the Tiptons and others were mad at the Forest Service for not completing the fence; the wildlife agents were angry that their trust had been abused. Accusations and letters flew in all directions.

Such was the climate in December when we all headed to Austin for the annual biological planning session. I left home long before dawn, and I remember speeding down the highway in the dark, contemplating what lay ahead. We had hired a facilitator for the meeting. I'd never worked with a facilitator before, but I imagined it as a sort of touchy-feely process. I figured he would start out building empathy among members of the group. Perhaps he would have each of us state the problem in our own words, and I set about articulating my view. Or perhaps he'd start by having each of us describe someone *else's* view of the problem. I tried to imagine how Tony and Jerrie might see it, or how Dayle Flanagan, the forest ranger, might state his case. But then I arrived at the meeting and Steve Rich, the facilitator, didn't invite us to talk about the problem at all. Instead he asked each of us to introduce ourselves, say why we had joined the team, describe the landscape when we first saw it, and describe it as we saw it presently.

At first, many of us were impatient. We didn't need introductions. We already knew one another; some of us knew one another so well we were hardly talking. We had too much to accomplish in too short a time; we needed to get down to business. But as we went around the table and listened, each of us reconnected with the authentic passion that had brought us to the team in the first place, and what had brought the others. We remembered that each of us was there because we wanted the land to heal. As we heard descriptions of changes in the land and recorded our own observations, we realized that, for all our setbacks, we had accomplished incredible things. Many of the people in the room had spent their lives trying to improve the health of western lands, and every person, every single one, said they had never before realized the success that we had been able to achieve. By the time we had completed the circle, we were no longer at war. We were a team again, trying to figure out a way to make things work. The truth of the land had united us, winning out over opinion and blame.

Cooperation, of course, is not a new story in the American West. If the negative side of individualism can take on an intractable, "get out of my face" quality that too often finds recourse in bureaucratic regulation or litigation, the belief that individuals can band together to solve their own problems underlies western traditions from early open-range roundups to community barn raisings to present-day volunteer fire departments. What stands out in the story of the Toiyabe Project is the complexity of that cooperation: how many different interests sat at the table together; how different we were from one another in our backgrounds, our dress, our politics, and even our patterns of speech; how we learned to find common ground through the very ground—the home land, if you will—that had brought us to the table in the first place.

It has been a decade since I worked with the Toiyabe group, and from this vantage point there is both good news and bad news. The bad news is that the Toiyabe Watershed and Wildlands Management Project did not survive. It did not fail for ecological reasons; by all objective measurements of ecosystem health, the watershed improved immensely. It did not fail for cultural reasons; rather, it united people with a wide variety of interests, agendas, and beliefs around a common goal. It failed for economic reasons. While we each came with our personal wish list for the landscape—increased bird habitat, trout habitat, wildlife habitat, abundant flowing streams and vegetated vistas—the cost of providing these things was financed entirely by the ranchers, Jerrie and Tony Tipton. With the exception of the incomplete Forest Service fence, they bore all the costs of capital improvement and daily operation. To keep cattle off the riparian areas, they hauled water, which required a massive outlay of both dollars and manpower; to keep cattle moving, they hired riders (and darn near worked themselves to death). They hosted the periodic monitoring sessions and team meetings; they launched an extensive education project to bring more people to the team and distribute its results. Because this model of collaboration was so new at the time, everything moved slowly—the fence, for instance, had been deemed necessary early on, but took years to move through the bureaucracy. In the end the costs were too high for the Tiptons to meet alone, and they had to give up the allotment. Effective restoration came to an end.

The good news is that the Toiyabe group and other early collaborations provided models and experience that have allowed virtually hundreds of other groups to form and thrive. The Tiptons went on to spearhead an even larger collaborative restoration project farther south in Nevada, this one financed by a creative blend of private, foundation, and agency funds, but that's just the tip of the iceberg. Just enter words like "collaboration" and "ranching" on Google to get a sense of what's out there, from the Malpai Group in the Southwest, which has succeeded in keeping over a million acres in open space, to Country Natural Beef in the Northwest, a consortium of more than 100 family ranches committed to raising healthy, environmentally responsible beef. Dozens of organizations such as the Quivira Coalition and the Sonoran Institute have formed to help bring people, land, information, and funding together. These are radical acts of cooperation that are redefining self-reliance and reaffirming the individual spirit.

There are still plenty of folks at war in the American West, but in their midst a new breed of warriors has arisen, people who are fighting to create what many call the radical center, a place where people can find common ground for the truly radical activity of healing both land and community, of creating a West that works.

I invite you to read on, to find out more about efforts to do just that. You'll find stories by people working to heal lands, foster collaboration and cooperation, and remind us what's worth preserving in our beautiful, star-swept West.

Richard L. Knight is professor of wildlife conservation at Colorado State University. In this essay, he suggests that open space and healthy food can be the focus of reconnecting rural and urban communities. First he examines this divide, and then he shows what people, both rural and urban, are doing to stitch it back together.

Bridging the Great Divide:

Reconnecting Rural and Urban Communities in the New West

Richard L. Knight

Enclosed is the Incidental Take Permit (Permit Number TE-115609) that The Nature Conservancy applied for in cooperation with the Livermore Area Landowner's Group to incidentally take the federally threatened Preble's meadow jumping mouse (Zapus hudsonius preblei) *in the Livermore Area . . .*

These were the opening words in a letter that my wife Heather and I—and our neighbors—received from the U.S. Fish and Wildlife Service. The letter was accompanied by instructions stating what we, as private landowners, were allowed to do on our land and what our obligations were to the federally threatened species that called our valley home. After six years of effort, we had received an Incidental Take Permit approving our Habitat Conservation Plan. In a nutshell, the federal government believed that we could coexist with a species that was in jeopardy along the length of the Colorado Front Range. Here in the Livermore Valley, at least, there would be peace on the range.

It didn't always look that simple. I vividly recalled the morning six years earlier when Heather told me that two Prebles mice had been live-trapped on our place. I was in Washington, D.C., teaching a week-long course on ecosystem management to Jamie Clark, then director

13

of the Fish and Wildlife Service. With her leadership staff, Jamie had asked my colleagues and me to explain the concepts behind collaborative conservation. When Heather and I spoke that morning, I never expected to hear the thrilling news that our place, so dearly loved by us, was also home for a species imperiled with extinction.

Shortly after returning home, however, reality set in. In the excitement of knowing we offered safe harbor to a species at risk, I had forgotten that the Endangered Species Act is one of many hot-button issues that divide the American West, and America, for that matter. Soon the residents of our valley became aware that we supported a large and thriving population of this nocturnal, winter-hibernating, streambank-loving, long-tailed mouse. The Livermore Valley is one of only a few ranching communities left along the Colorado Front Range, an area better known for its 160-mile-long expanse of annexing cities, exurban sprawl, traffic congestion, brown cloud, and seemingly endless shopping opportunities. The ranching families that called Livermore home kept the valley in open space while also husbanding livestock and stewarding land.

And now we would have to deal with the threatened mouse, a bugaboo that has bedeviled so many other western communities that deal with the requirements of the Endangered Species Act. Like ham and eggs for breakfast, the chicken might be interested in what's for breakfast but the pig is committed. Would it be possible, to paraphrase Aldo Leopold, to live in our valley without spoiling it?

Heather and I had been working with our neighbors on conservation-related issues for some time. Would our coalition of partners and landowners be able to withstand the environmental storm about to trouble our waters? In our hearts we thought so. After all, most of our neighbors agreed that conservation that works is conservation that works not only for natural communities but for human communities as well. Actions that benefit one at the expense of the other are not conservation, they are something else. On the one hand, if human communities grow weaker in order for natural communities to thrive, who will fill the role of land steward in the years to come? On the other hand, if the land is degraded and weakened, how long can human economies thrive on a dying landscape? Aldo Leopold understood this:

"Conservation means harmony between men and land. When land does well for its owner, and the owner does well by his land; when both end up better by reason of their partnership, we have conservation. When one or the other grows poorer, we do not."

Our neighbors and my wife and I will deal with the mouse issue as we have with other threats to our community, whether it be the enlarging of a dam, the paving of a rural road, or the loss of a ranch to subdivision. By going slowly and taking the time to understand all sides of an issue, and then reaching out to the good intentions that are a part of all human constitutions, we will seek common ground rather than focus on the things that divide us. This is home, the only home we may ever have. Though achieving cooperation can be as slow as watching glaciers melt, we choose to build bridges rather than erect barriers. The results are more difficult to accomplish, but our neighbors believe they are longer-lasting.

Our story is one of many that might be used to illustrate the dichotomy between the rural and urban West. City people want rural landowners to protect wildlife habitat, open space, and ecosystem services while landowners feel that city people take for granted these societal benefits, without so much as a thankful nod to their rural neighbors. Meanwhile, the economic reality is that their efforts to produce food and fiber are increasingly placed at risk by our global economy and by cheap food from offshore. The rift between the West's rural and urban societies can be overcome only when we appreciate what each produces and the natural interdependencies that bind us.

A West at War

If the message that the fates of humans, land, water, plants, and animals are entwined and indivisible is obvious to so many of us, then why do so many fracture lines exist today across America and, especially, in the American West? Why is the West at war with itself?

Perhaps we should begin with an easier question. Why is the West seemingly at war with the rest of America? On any given day the major media outlets issue some story that slants its message of westerners besieged over a natural resource issue. The underlying theme in most of these stories is that all of America will lose something of great value if westerners are allowed to exploit some resource. Whether it's the fate of wolves in the arid Southwest or that of snowmobilers in the northern Rocky Mountains, Americans outside the West fret over the health of the western landscape far more than westerners worry over American lands in other regions.

To explain this phenomenon, consult a geographer. The American West, unlike other parts of the United States, is evenly divided between public land and private land. When Americans wake up each morning, they do so knowing that half of the West is their birthright. Although they comprise distant communities of interest, they do understand the guarantees that come with living under a federal form of government. Half of the West is public, owned by and open to all Americans.

Because of this public-private land dichotomy, there will always be tension between westerners and the rest of America. So long as federalism holds sway and does not lose ground to devolution by the New Federalists, Americans will have to deal with the concept of a federal domain. In the meantime, being a bioregionalist, I am concerned primarily with the clashes between communities of interest and communities of place *within* the West. The Endangered Species Act is just one issue that ignites controversy in the West. Like America at large, other issues separate us, whether it be gun control, same-sex marriage, or pro-life/pro-choice struggles.

The Red, the Blue, and the Purple
In confronting the West's dilemma, we are all but compelled to look upon the now-infamous maps of Red-Blue America. During the 2000 and 2004 presidential elections, America seemed to divide into two camps: a Red Republican group and a Blue Democratic group. The West, other than the three Pacific Rim states of Washington, Oregon, and California, appears as a solid polygon of scarlet red. Although political pundits have made much of the Red-Blue *states*, it is more interesting to look at Red-Blue *counties*. The three western states that went Blue, when examined at the county level, quickly take on the color Red. Most metropolitan counties in the West are true-blue Democratic; the rural western counties are Republican. This rural-urban divide mirrors the division between ranchers and farmers in the

rural West and the region's majority population, anchored in urban and suburban centers. Red predominates when it comes to acreage; Blue triumphs when it comes to population.

A more careful look at the Red-Blue map reveals further subtleties. Demographers provide us with a third color when examining the rural-urban, liberal-conservative divide: purple. When you mix red and blue, you get purple. On the map, Purple demonstrates that the "urban" cores may be Blue, but as you leave downtown and ease into the suburbs, you come into increasingly Republican neighborhoods.

Just as suburbs complicate the term "urban," exurbs make it increasingly difficult to neatly categorize "rural." For the first time in over a century there are more Americans leaving metropolitan areas to live in the country than there are rural people moving to the cities. Americans, always consumers of good places, are buying small acreages on what had only recently been farms and ranches and living exurban existences on ranchettes. Demographers might say, therefore, that the rural-urban divide is passé; today's fault line is more appropriately called the suburban-exurban split. Having worked our way outward from urban to suburban, and now to exurban, one wonders where we will settle next. In gated communities?

I shine light on this only to ensure that the true complexities of our contemporary rural-urban relationships are not overly simplified. Though Red-Blue, Rural-Urban, Conservative-Liberal, Democratic-Republican dichotomies are orderly and understandable, the changing faces of the New West are not so neat. Interestingly, whether city or country, the fundamental disconnect in the rural-urban divide is still that between people who at heart think *city* when they arise in the morning, whether in a suburb or on a ranchette, and ranchers and farmers who think *country*, whether still on the land or now in an urban retirement home.

Food and Open Space: Stitching Together the Rural-Urban Divide

Despite misguided attempts to create rural-urban dichotomies, these alleged differences are, at their core, false. All westerners prefer healthy food and prize open space. Farmers and ranchers produce both though they are compensated only for the first. That more and more of our food comes from offshore, and that we are content to shop at box stores for the best cut of meat for the lowest price, makes the job of agriculturists, who guard the West's open spaces, more tenuous. Likewise, agriculturists need to acknowledge the importance of cities. This is where their customers are; where the libraries, hospitals, universities, colleges, and county courthouses are; where the retail businesses are; and where the factories that build pickup trucks and horse trailers are. If there are true differences between rural and urban in the American West, they reside in a bubble that will soon burst. All of us would have to exist in a state of extreme denial to believe that the fates of rural and urban communities are not inherently entwined.

One answer to our dilemma is to connect local food production to local open space, economically and ecologically. This will require that westerners eat locally produced food that is well husbanded on land that is well stewarded. To dismiss this solution as nostalgic misses the point. We know that the earth's soil, water, atmosphere, plants, and animals cannot sustain food production as presently practiced under the agribusiness model. To accept an alternative model of agrarianism is to quicken the day when western communities once again realize their interrelatedness: locally produced food and open space offered and received with grace and a fair market value by urban people who no longer take for granted the societal services of local farmers and ranchers. If the truth of these beliefs requires referencing a "higher authority," then visit the Bible: "That which happens to men also happens to animals; and one thing happens to them both: as one dies so dies the other, for they share the same breath; and man has no preeminence above an animal: for all is vanity."

Wendell Berry sank the stake when he wrote:

The most tragic conflict in the history of conservation is that between environmentalists and the farmers and ranchers. It is tragic because it is unnecessary. There is no irresolvable conflict here, but the conflict that exists can be resolved only on the basis of a common understanding of good practice. Here again we need to study and foster working models: farms and ranches that are knowledgeably striving to bring economic practice into line with ecological reality, and local food economies in which consumers conscientiously support the best land stewardship.

Good news abounds regarding open space and local food production. There is evidence aplenty that urban and rural people are rediscovering their historic connections. Urban people line up in droves to approve ballot initiatives that protect open space, thumbing their noses at elected officials who fail to see the high esteem that urban people place on open lands. From 1998 to 2003, voters passed 76 percent of 802 conservation ballot measures, generating $24 billion for protecting open space (www.landvote.org). In 2004 alone, 217 initiatives were passed, generating $4 billion for keeping land undeveloped.

The number of land trusts holding conservation easements is keeping apace with dollars for open space protection. In 1981 there were 370 land trusts in the United States; today there are more than 1,400 (www.lta.org)!

Even ranchers and farmers are getting in on the act, proving beyond a doubt that they also would rather farm and ranch than sell out to a developer. In Colorado, the statewide cattlemen's association has created the Colorado Cattlemen's Agricultural Land Trust (to whom the proceeds of this book are committed). The first land trust in the nation to be formed by a group of mainstream agricultural producers, they now hold easements on over 200,000 acres of private agricultural lands. As demonstration that this sentiment among western ranchers and farmers is regionwide, six other state cattlemen's associations now have land trusts for agricultural producers (www.maintain therange.com).

Ranch families have expressed the ultimate private property right, denying forever the "right" to develop lands that are the most productive in the West, the best watered lands with the deepest soils. These ranch and farm families have drawn a line in the soil that speaks to their commitment to land, family, community, and the production of homegrown food. In voluntarily conceding forever the last cash crop of many western farms and ranches, these individuals illustrate the differences between "takers" and "caretakers." As my neighbors Even and

Catherine Roberts told me years ago, they never felt as if they *owned* the land; they felt as if they *belonged* to it.

Agriculturists are taking their message to town as well, demonstrating that producing food for local communities is good for the bottom line. These risk-taking individuals can be found almost everywhere, though they are clearly not the dominant voice of western agriculture—yet. By stepping out of the "agribusiness model," they are symbolic of what can be, and they help to demonstrate that ranching and farming, done right, can be culturally robust, economically viable, and ecologically sustainable. Farmers' markets are doubling every ten years. For example in 1994, America had 1,646 markets selling locally produced food; ten years later the number had swelled to 3,700 (www.farmersmarkets.net).

Look anywhere in urban America for evidence that urban people and agriculturists are both doing their share in forging an intelligent consumption-production equation. The organic revolution is on. Organic products are sold in nearly 20,000 natural food stores and cooperatives and in nearly 73 percent of conventional grocery stores. The U.S. organic market is expected to grow over 21 percent from 2002 to 2007, reaching a value of over $31 billion. I can't help but be touched to hear how students at my home institution of Colorado State University are hearkening back to the century-old land grant message in ensuring that what students eat in our dormitories comes from local farmers.

Montana is tapping into this slow food movement, ranking first nationwide in organic wheat production, and second in production of other grains, peas, lentils, and flax. Working through companies like Great Harvest and Wheat Montana, Montana farmers are growing healthy food locally and selling it locally. The same trend is emerging for grass-fed beef. Just ask Andrea and Tony Malmberg, Wyoming ranchers. By producing grass-finished beef, they are allowing consumers and restaurants in Wyoming's cities to eat a healthy and safe product produced on Wyoming's open spaces, both private and public lands.

These trends are all growing and point in the right direction. They capture the necessity that David Orr wrote about in *The Last Refuge*: "For agrarianism to work, it must have urban allies, urban farms, and urban restaurants patronized by people who love good food responsibly and artfully grown. ... It must have farmers who regard themselves as trustees of the land that is to be passed on in health to future generations. It must have communities that value farms, farmland, and open spaces." Perhaps we should take a rancher or farmer to lunch and discuss food and open space, and human and land health.

Rangelands: Blended, Half Public, Half Private

When dealing with rural-land relations it certainly pays to keep an open mind. Ignoring our interdependencies only accelerates the rural land-use conversion that is sweeping across the West, with a realtor running through it. At one end of this land-use spectrum stands a farmer or rancher, and at the other end a developer. The tensions are not eased when well-intentioned pundits issue inflammatory declarations. Thomas Michael Power, an economist, wrote in his book *Lost Landscapes and Failed Economies* that the

> primary environmental objection to expanded residential activity is that subdivisions and urbanization damage the landscape in a variety of ways. This no doubt is true if subdivisions are compared with, say, protecting land as a wildlife preserve or as wilderness. But that is rarely the alternative use to which the land would be put. The appropriate comparison is between the environmental impact of ranching activity and that of residential use. We must put our agrarian sympathies aside: ranching does not step lightly on the land.

This is a profoundly chilling statement. When a learned westerner can write something like this, it should be a wakeup call to all Americans. Amazingly, although exurban development is the leading cause behind the decline of federally threatened and endangered species in the continental United States, statements such as Power's are more the norm than the exception.

If urban people value ranchers and farmers primarily for the open space they protect, then shouldn't we be concerned about the relationship of this private land to the half of the West that is public?

The approximately 31,000 grazing leases on BLM and Forest Service lands are connected to over 107 million acres of private land on which ranchers graze their sheep and cattle the rest of the year. What would happen to wildlife and open space if public land grazing were to end and the private lands were developed? Public lands, being less productive, cannot sustain healthy wildlife populations once the private lands rimming their boundaries are developed and reappear as subdivisions. Yes, magpies and raccoons will do just fine, but not bobcats and yellow warblers.

Colleagues and I recently examined plant and songbird communities on ranchlands, protected areas without grazing, and exurban ranchettes. Populations of common species, such as robins and magpies, and dogs and cats, thrived amid the ranchettes. The ranches and protected areas supported populations of lesser-known species, such as Brewer's sparrows and towhees. Ranches had the most intact grasslands with the greatest cover by native species, the fewest number of weeds, and the least amount of bare ground. Regretfully, the protected areas and ranchettes were the weediest. Weeds aren't good for the bottom line of ranchers. But ranchers have a great management tool to control invasive species: the judicious use of livestock. Public rangeland without livestock may be appealing to some, but will we have the land stewards to ensure that weeds are managed? Downsizing the federal workforce is increasingly leaving our public lands abandoned and turning feral. Ranchers, who steward their own lands, can also serve the role of custodian on public lands where their livestock graze.

What about the argument that livestock grazing is subsidized on public lands? It is, but then what use isn't? Indeed, outdoor recreation is the most subsidized use of public lands. Sadly, it is also the second leading cause for the decline of federally listed species. Only water development projects trump recreation. But then, if our conversations were honest, we wouldn't call any of these public land uses "subsidized." They are allocations of your tax dollars to promote authorized uses of federal lands. After all, owners of snowmobile and off-road vehicle businesses whose livelihoods depend on access to public lands view their clients' use of western public lands as both worthwhile and contributing to local economies. And how many acres of productive private lands do motorized recreationists keep out of development?

This needs to be said. Anti-grazing extremists fail to understand that for every piece of public land grazed, there is also a piece of private land committed to ranching rather than residential development, and that given today's land prices, there is no other alternative but to sell the deeded land when the grazing permit is canceled.

What does this have to do with ranching and the latest attempt to end all public land grazing? Those individuals who champion the end of grazing on public lands are aiding and abetting the cause of developers. Indeed, it could be argued that these individuals are as instrumental in the development of the half of the West that is in private ownership as are our Chambers of Commerce and most ardent New Federalists.

What if anti-cow environmentalists expended their energies working with ranchers to ensure more appropriate livestock grazing? Might we actually make peace on the range and better steward the land, both private and public? Wouldn't that be a better outcome?

There are many things the West needs right now, but there is nothing it needs more than healthy rural lands supporting healthy rural communities with the full awareness and commitment of those living in healthy cities. That ranching will survive in the New West is not the question. The question is whether ranching will rediscover its ability to once more encourage vibrant connections between urban and rural peoples, focusing on the rural lands of the West that bind us all to this remarkable region.

Home, Land, and Security

During the heat of the political campaigns in 2004, Americans may have missed two pronouncements. First, the Department of Agriculture briefly mentioned that for the first time Americans imported more food than we exported. Second, outgoing secretary of Health and Human Services, Tommy Thompson, warned that terrorists could easily reach across our oceans and hurt Americans through our imported food.

That we now buy more food from offshore than we grow onshore is the inevitable consequence of our globalizing economy. Tommy

Thompson's code orange alert was quickly pooh-poohed by the White House. All is well in the American homeland. Or is it? And what do these bits of old news have to do with the American West? David Orr wrote in *The Last Refuge*, "The resilience that once characterized a distributed network of millions of small farms serving local and regional markets made it invulnerable to almost any conceivable external threat, to say nothing of the other human and social benefits that come from communities organized around prosperous farms."

Clearly our politicians don't get it. The political campaigns of recent years have been full of rhetoric over manufacturing jobs going offshore with nary a word over the loss of agricultural jobs to other countries. The covenant that once bound rural and urban Americans has been severed. Until recently, urban people knew where their food came from, and supported it. Rural people, in return, provided open space and places for city people to hunt, fish, and visit.

In this light our government's concern over "Homeland Security" misses the most important point, and is little more than overheated rhetoric from politicians who typify the type of individual that Samuel Johnson described when he wrote, "Patriotism is the last refuge of a scoundrel." A secure homeland is not simply based on military might. Home, land, and security also come together when urban people realize that ecologically sustainable food production is possible and that rural cultures matter, and when urban people are prepared to compensate farmers and ranchers for a healthy food product as well as for protecting open space, wildlife habitat, and watersheds through husbandry and stewardship. Equally important to this winning equation are rural people who acknowledge the importance of urban areas and offer a friendly handshake to their urban neighbors.

To find evidence that this radical centrist position is tenable is to look at almost any watershed today. In the words of the "Radical Center" group that first met in New Mexico in 2003 (www.quivira coalition.org):

> We have two choices before us. One is to continue the heated rhetoric of the far right and the far left, spending our time slinging insults and hardening polarization. Or, we can join with the

hundreds of watershed and community-based programs around the country and move to the radical center. The point where people will respectfully listen, respectfully disagree, and in the end, find common ground to promote sound communities, viable economies and healthy landscapes. By following this behavior we will, over time, so marginalize the wingnuts on the right and left that our voices will be heard over theirs.

To return to reality in the New West will require what Aldo Leopold warned us about in 1928:

Even the thinking citizen is too apt to assume that his only power as a conservationist lies in his vote. Such an assumption is wrong. At least an equal power lies in his daily thought, speech, and action. ... But most problems of good citizenship in these days seem to resolve themselves into just that. Good citizenship is the only effective patriotism, and patriotism requires less and less of making the eagle scream, but more and more of making him think.

For security's sake, it's time to beat those red and blue swords into purple plowshares!

In this essay, Diane Josephy Peavey writes about her life on a sheep and cattle ranch in south-central Idaho, incorporating into her personal story its history, people, and the changing landscape. She began writing, and then organizing the community, to celebrate this life and its stories when its future was threatened first by the farm depression of the 1980s and then by environmental hardliners who wanted family ranchers off the land.

Confronting Fear

Diane Josephy Peavey

The radical center became a sanctuary for my husband and me, as ranchers and environmentalists, when we suddenly found ourselves living several miles away from the headquarters of the Western Watersheds organization, a vociferous group of intractable ranching critics. They quickly became a center for environmental extremism espousing the end of public lands grazing. Hostility flared up between neighbors and radical environmentalists as letters to the editor and newspaper articles vilified those of us living on, and caring for, the land and western open space. The end of ranching wouldn't cause an "economic hiccup," Watersheds director and local architect Jon Marvel told the press. On another occasion he generously conceded to an environmental gathering that, although he planned to get rid of ranchers, "I'll let them keep their music."

Former friends who had worked with my husband and me on conservation issues from water quality to nuclear waste disposal suddenly began to question our ranching practices, of which they knew little. And although I had always been proud of my husband's care and his strong connection to the land, I was suddenly unsure I could explain our lives to these friends turned critics. They were no longer listening.

Home became a new and frightening place.

So it was a relief to visit the Malpai Borderlands Group in New Mexico and learn about their concept of the radical center, a place to work through western land conflicts. Here calm prevailed and people of sensible minds and sincere passions gathered to save threatened landscapes. I no longer felt alone.

Since then, I have turned to the radical center often but only recently found a way to remain there, thanks to a band of sheep, a bike path controversy, and my husband's eminently reasonable conviction that if we share our ranching lives with those who know little about living on the land, even those hostile to us, we will make friends for rural and ranching communities.

When I moved to a sheep and cattle ranch in Idaho directly from urban America over twenty five years ago, I had little understanding of the life and certainly no idea of the deep human connection to landscape that exists among families of the land. It took years of experience before I found myself comfortable here and then, finally, tied inexorably to this life.

My early reticence was a result of timing. I arrived in the early 1980s at the start of a farm depression that rivaled the one in the 1930s. Life was not easy, and in the middle of agonizing turmoil, foreclosures, and bankruptcies, I watched dazed neighbors and friends lose everything and leave the land in staggering numbers. The exodus was triggered by severe drought but more insidiously by plummeting market prices manipulated by corporate agricultural conglomerates. It was a painful introduction to ranching, and I weighed this new life wondering if I really should adopt this desperate place.

So I watched. Some family ranchers and farmers were able to hang on for dear life. They argued with naysayers and pleaded with bankers to stay with them until prices for their crops and animals pulled out of their crippling slump. They schemed, diversified, struggled to survive. Many didn't make it.

Those who did, like our family, took a deep breath and moved forward, still shaky, and yearning for better years. But almost immediately we ran headlong into the next wave of adversity, the fight over the use of public lands framed and forwarded by radical environmentalists led by the Western Watersheds group. We were caught off guard by the venom they brought to the landscape, unusual among environmentalists with whom we had worked in the past. These hardliners marched to their own belligerent chant, "Cow Free by '93."

Worse than their chilling slogan, it became clear they were uncompromising in their goals, unwilling to talk or come to our ranches and walk the land and work with us. Many of us offered an invitation but were ignored or rejected.

They meant to get rid of us, all of us in the arid West who raised livestock. And they were going for the jugular. Without the ability to graze public lands in summer when we were growing hay for winter on our private lands, many families would not have enough feed for their animals and would be forced out of business.

Indeed, as a public policy, "cattle free" would sound the death knell for family ranches still teetering perilously on an economic ledge. It would also jeopardize the care and sustainability of western lands. These are enhanced by seasonal grazing, which controls the growth of grasses susceptible to fire and replenishes those same grasses when dried seeds are pushed into the earth under animal hooves in the fall.

Clearly there were deep pockets of misunderstanding around the West. It seemed to be an unsettling consequence of ranching, and I wondered if life on the land would ever be calm.

But at the same time, I began to understand that our case was aggravated by our ranch's proximity to Sun Valley. This famed ski area began in the mid-1930s when the Union Pacific Railroad created a resort in the heart of sheep-ranching country. Over the years it became the playground for the rich and famous: Gary Cooper, Marilyn Monroe, John Kennedy, and Ernest Hemingway, among others. But since the 1980s the area has expanded beyond the resort center, and part-time residents vacation lavishly in second homes along green golf courses throughout much of the long and lanky Wood River Valley. The area has become the quintessential picture of the New West with new populations, new housing, and new upscale services.

When I first visited the area, it was still surrounded by family ranches. I met and married John Peavey, a third-generation Idaho sheep and cattle rancher who lived and ran livestock in the foothills of the Pioneer Mountains. At the time, he was also a state senator, a local hero. John had helped found a statewide environmental organization in the 1970s and been a prominent spokesman for conservation issues in the state senate and independently throughout the state. He led successful campaigns to save southern Idaho's remarkably clear air from the pollution of a proposed coal-fired plant. He worked to protect the waters of the Snake River from rapid development and overappropriation. He sounded the alarm on the unchecked aquifer contamination coming from Idaho's sacrosanct nuclear facility.

John's grandfather had gotten into the sheep business in the late 1920s, but earlier in the century others had led the way. The country looks much like Scotland, I'm told, and many of the original Scots arrived from the old country to work on western sheep outfits while they built up their own flocks. They eventually started some of the large ranches that still operate in the region today. The animals took well to the yearly migration from winter desert pastures to summer mountain-meadow grazing. The families and numbers of sheep grew until soon after World War I, it is said, the area was second only to Sydney, Australia, as a sheep center.

Today our sheep still move through this country, but things have changed as a result of years of discouraging lamb and wool prices, pressures from corporate agriculture, and pressure by developers to buy our private lands at astronomical prices. Instead of hundreds of thousands of animals and many ranching families traveling the valley's historic trails, only about 30,000 sheep run by five outfits remain. And although for the moment our route is presumably secured by traditional easement guarantees, many pastures along the trails are closed off by new development and recreation that keep us on a narrow path. It is that way around the West.

So into the mix of rapid growth and dwindling sheep ranches came the new brand of environmentalist aimed at removing generations-old ranching families.

We watched Jon Marvel disrupt these ranching operations by out-bidding ranch owners on their state land leases. He claimed that low lease prices—high to struggling ranchers—were a misuse of state monies that should be earning more for the public school endowment fund. This reasoning held a certain cachet with urban supporters and valley newcomers who knew little about ranching and the important role ranching serves in keeping open spaces out of development. So with a ready cash supply Marvel bid high enough to push lease prices out of reach of many ranchers. But these machinations paled next to his skills at handling the press in order to turn neighbor against neighbor.

Often in cruel language, he belittled old-timers and young ranching families. His disparaging remarks were strung across newspaper and magazine pages like those in the 2004 winter issue of *Range* magazine when he claimed, "ranching is an atavistic lifestyle, an unusual relic of the past that has nothing to do with reality. It's not financially viable so how can it be a business. They [ranchers] need to sell their land and get jobs like everybody else."

In the same article, Oregon rancher Michael Hanley remembered: "The first time I met him, he [Marvel] came up and said, 'I'm going to do whatever it takes to get rid of you and your cows.' He got close to my face and poked his finger into my shoulder. He wasn't laughing. Now how do you and I talk to someone like that?" Hanley asked.

I have wondered the same thing, especially when I find him everywhere in our lives just down the road, mixed up with friends at meetings and social gatherings. For years my heart raced in panic when I saw him. The hate seemed unending, impenetrable, and it continues today.

Recently Idaho signed a wolf management plan with the Department of Interior that turned federal control of these animals over to the state. Instantly the hue and cry from hardliners went forth in newspapers statewide, charging that ranchers now will be killing wolves indiscriminately—without provocation. "Who says so?" ranchers ask. But no one responds and the seeds of slander are sown before the ink dries on these agreements.

Yet despite the fear, many of us around the West remain convinced that understanding will subdue intolerance. And so we look for ways to reach out, to tell our stories.

At our ranch we began to market our lamb locally at restaurants and markets to build appreciation for our work. What better way to make friends than a good rack of lamb from animals raised in high mountain meadows? But the work proved highly labor-intensive for our crew, already stretched thin.

About the same time an unusual occurrence unfolded. The county asked local sheep producers if they would agree to a county bike path along the traditional stock route that stretched the length of the Wood River Valley. "Not a problem," the ranchers responded in their usual neighborly way. But "not a problem" became one as soon as the concrete dried and the first band of sheep moved through the valley. The calls began.

"The sheep droppings are getting caught in my Rollerblades and my bike tires. What are those animals doing on *our* bike path? It's disgusting."

Disgusting? Sheep? Rollerblades? Wait a minute. This was an easement reserved for the migration of sheep and used for over 100 years. Were historic rights once again to be pushed aside when their promises became inconvenient?

We paid attention to these calls as they increased in number and discontent, and soon my husband came up with a plan. Why not invite the community to help us herd the sheep through the valley? Certainly we could keep the sheep off the bike path and, even more important, share ranch life with this aggrieved group. Why not?

So in a newspaper ad we invited people to meet us for coffee at a small café at 6 A.M. There, over steaming cups of coffee, my husband talked about the sheep business, his family's seventy-five-year history moving the animals through the valley, raising them on native grasses and plants. He told stories about the dogs, the sheep camps, the herders—at one time all Basque but now mostly Peruvians. He explained what made good feed and safe conditions for the animals and about the migratory life of sheep ranching.

Then, with first light of day, the group of twenty would-be herders pulled on warm gloves, jackets, and hats and stepped into the cold to catch up with the 1,500 sheep. And all of us began the move south, *along the bike path,* that mid-October morning.

To our surprise this informal outreach became popular, growing yearly in size and enthusiasm. Soon friends were joined by the curious and by classes of schoolchildren throughout the day. The event became an anticipated fall occasion.

The Sun Valley–Ketchum Chamber of Commerce took notice and suggested we discuss turning our small effort into a larger program. They would help.

We met in April with representatives from the valley's chambers, libraries, museums, college extension office, University of Idaho county extension service, the school district, and other local groups. The interest was inspiring. Perhaps family ranchers still had friends after all.

That October, 1997, we launched the Trailing of the Sheep Festival, not a reenactment of history but history in the making. With or without weekend events, ranchers would be moving their sheep south out of the high mountains before the winter snows, as always. But now we held a huge parade of 1,500 sheep to celebrate this history. And although the animals were perhaps a little skittish in front of crowds, they performed heroically, running, jumping, backing up, racing forward, and spinning down Main Street.

This was Ketchum, Idaho, the resort community, with its small boutiques, flashy restaurants, and glitzy ski-wear shops. This was the Wood River Valley, the heart of the anti-ranching movement. And these were sheep on parade. Oh yes, the first year, there was a lone demonstrator with a stenciled sign claiming environmental degradation by sheep. To our amazement he was booed all the way through town.

The 1,500 sheep were the stars, followed proudly by sheepherders, dogs, and camp tenders pulling the camp wagons. And true to our earlier event, if visitors wanted to become herders, they could fall in behind the animals and help move them to their afternoon resting spot several miles beyond the parade route.

Now the festival weekend offers up one evening of sheep readings and music, and a second of sheep tales filled with reminiscences of the old days of sheep ranching in Idaho. There is a daylong Sheep Folklife Fair with the music and dance of the people who figured prominently in the sheep business, the Scots, the Basque, and the Peruvians. There is a Basque lamb dinner, demonstrations of spinning, weaving, sheep shearing, and sheepherding with dogs.

In this, its tenth year, we expect 10,000 people will fill the town and participate in the event. The weekend has sparked the national imagination and brought a new pride to the community. There has been press coverage from *Business Week* to *Sunset* magazine and stories in papers from the *Los Angeles Times* to the *New York Times* and most points in between. The 2001 festival was filmed by *CBS Sunday Morning* as a kind of national antidote to the 9/11 tragedy.

Those of us who remain in sheep ranching still reel when we read angry letters in the paper claiming that the sheep cause environmental damage. We know better. We know the importance of grazing for the health of our lands, and we know that Americans now more than ever celebrate natural foods like our succulent lamb raised in the mountains of Idaho.

From the beginning the festival was a kind of challenge to the community to learn about the history and culture of its landscape, to listen to the stories of this place they call home. And they have done just that.

Ranching in the radical center for us means sharing our life with those who want to know what we do and why. So we take the curious and possibly the hostile head on, telling a story so basic to survival, so intrinsic to the land and its history, that it has smoothed rough edges, quieted brashness, and offered a promise of hope for this western country.

And best of all, we've been told that old-time sheep men have come to the Trailing of the Sheep to witness for themselves this remarkable event. Some are mystified by all the fuss, while others have been seen to brush a tear away when people they don't know, crowds of them in fact, cheer and applaud a band of ordinary sheep and a lifetime of hard work unexpectedly celebrated and honored.

Paul Zarzyski is the author of several volumes of poetry and has read at the Library of Congress and at the National Cowboy Poetry Gathering in Elko, Nevada. Subtle memories have more to do with our love of place than we imagine. With this poem Paul Zarzyski celebrates the sweet earthiness of a lucky life.

One Sweet Evening Just This Year

Paul Zarzyski

Sundown rolling up its softest nap
of autumn light over the foothills, grass
bales stacked two tiers above the '69 Ford cab,
our long-toothed shadow slices east,
mudflaps dragging dry gumbo ruts
back home after one beer
at the Buckhorn Bar quenched the best
thirst I've worked-up
all millennium, pool balls
clacking above the solemn
cowmen reminiscing their scripture,
waxing poetic lines to The Legend
of Boastful Bill—*one sweet morning
long ago*, the hands-down favorite. I'll bet
this whole load, that old bard,
Charles Badger Clark, knew the eternal
bent of those words
the instant he scratched them across the open
range of the blank page.
                    Glacial melt
runneling over mountain rock,
moist air swirls in the cab

stirring up three decades of Montana
essences atomized
into a single mist, this horse-cow-dog-grit-
gunpowder-drought-leather-sage-sweat-
smoke-loss-whiskey-romance-song
fragrance settling upon the porous
inner wrist of dusk
unfolding for only a moment
its sweet, unique blossom.
                                           And me, tonight
I'm the lucky one along for the ride,
head still sweaty beneath my hat,
a harlequin glitter of hayseed
sticking to my bare arm stretched straight
out the window for no reason
but to know my own pores rising
beneath hair pressed flat
and flowing like grass in crick-bend shallows,
timothy in the side mirror, stems hanging on
with one arm and waving
wild with the other—to golden meadows
and rolling prairie flecked with cattle,
antelope, jackrabbit, grouse,
all grazing beneath one big gray
kite of bunched starlings'
acrobatic flashings over stubble.
                                           We mosey home,
me and the old truck, in love
with our jag of good Montana grass—
not one speck of simplistic myth
between us and the West that was, sometimes
still is, and thus will be
forever and ever, amen.

—*for Ralph Beer and Wallace McRae*

Nathan Sayre has worked on ranches as well as national parks, forests, and wildlife refuges. A student of history and geography, he tries to recover the human and natural processes that have produced the landscapes we encounter in the present. Here he considers the relationship of western cities to ranchlands over the past century.

The Western Range:

A Leaking Lifeboat for Conservation in the New West

Nathan F. Sayre

On the first Saturday in May 1960, the citizens of Tucson awoke to learn that their city had agreed to buy a ranch. The deal was engulfed by scandal and nearly collapsed on Tuesday, when the city discovered that one of its negotiators had secretly acted as a middleman and stood to reap a windfall of nearly $300,000. On Thursday, under mounting pressure, he and his silent partner forfeited their option, allowing the city to void its original contract and buy the ranch directly from the owners. By the end of the month the deal was done and the scandal had passed, but in retrospect it was an omen. Both the city and its new neighbors soon regretted the deal, which marked the beginning of a half century of shifting and contradictory relations between urban growth and ranching in the region.

The ranch was thirty miles east of the city, over the Rincon Mountains in the neighboring San Pedro Valley. It included 2,200 acres of private land on the San Pedro River that the newspapers called "water-bearing lands." According to engineers, wells drilled there could yield fifteen million gallons of groundwater per day, which city officials planned to pump over the divide to supply the booming urban population. But the plan never materialized. Some farmers sued to stop a similar project west of Tucson, and in 1969 the state supreme court

ruled interbasin water transfers unconstitutional. In 1985 the city sold
the ranch in a sealed-bid auction to a developer, who divided the private land into forty-acre parcels arrayed around a gravel runway for
small aircraft. The development failed too—there was no paved road
access, and sales were slow—but there was no going back. In such small
pieces, the land was too expensive for livestock production. A decade
later, priorities had shifted again. As though suffering from seller's re-
morse, the city returned to the San Pedro in 1998 and bought the ranch
next door—*to keep it from being subdivided.*

Tucson's relationship to the San Pedro Valley may seem schizo-
phrenic, but it is far from unusual. One might even call it representa-
tive of the American West as a whole. Throughout the region, cities are
scrambling for water. Ranchlands are being converted to residential uses,
even in places remote from large cities. For the first time since the
Depression, more people are moving to rural areas than away from
them, not to raise crops or livestock but to experience a kind of rural
lifestyle: more space, more stars, more nature, and fewer people.
Whether they buy five acres or fifty, these "exurbanites" can afford to
pay more per acre than livestock production can justify. Development
follows the private land: up the middle of major valleys, out along trib-
utaries into the foothills. The census estimates that about one million
acres of farm- and ranchlands in the eight interior western states have
gone out of agriculture *per year* since 1964. In response, urban con-
stituencies are calling on the government to rein in the sprawl and protect the West's legendary open spaces.

In a matter of decades, the meaning of conservation in much of the
West has done an about-face, moving from protecting public lands from
livestock grazing to protecting private lands from development by keeping them in ranching. Across the region, groups of ranchers are organizing to protect their shared landscapes and watersheds, often in alliance
with conservation groups. Their efforts confront a landscape uniquely
fraught with both great opportunities and acute contradictions. On the
one hand, a significant amount of the West is owned by the American
public and is thereby protected from most (if not all) forms of development. National parks, forests, wildlife refuges, and holdings of the
Bureau of Land Management and the Department of Defense together

make up fully half of the land in the eleven western states. Indian reservations and state lands comprise another 11 percent. The vast majority of these lands have never been plowed or paved, giving them unique biological values and making it easy to equate "open" with "natural" or "wild."

On the other hand, the West's generous share of public lands means that development is confined to a narrower land base than would be the case in, say, New York or Georgia or Iowa. The rest of the country (excluding Alaska) is almost 90 percent private land. By contrast, Nevada and Arizona are the fastest growing states in the nation, but they are only 11 and 16 percent private land, respectively. All that demand for real estate is thus focused on an unusually limited supply. Expansive vistas make even modest developments conspicuous from great distances, so the western experience of vast open spaces can be dashed by the conversion of just a small slice of the landscape. Paradoxically, the West's open spaces are vulnerable because so much of the land is protected.

Why have these contradictions emerged in the past half century? What kept western rural private lands so open until so recently? How did we fail to notice them, intermingled throughout those famously open landscapes? The attention paid to public lands is understandable, given their beauty and the opportunities they afford to tens of millions of visitors each year. They constitute a unique American heritage, not to mention an extraordinary asset for tourism, conservation, and recreation. But the public lands are everywhere embedded in a matrix of private lands; both types are present in most every panorama we might choose to behold. We don't tend to notice the private lands until they are developed, but they've been there as long as the public lands, and in many cases longer.

The reason we could overlook private lands for so long was that a system of laws and policies formulated over a century ago linked their use to the use of the surrounding public lands, tying them together in such a way that the plat lines between them remained effectively irrelevant. Call this system the Western Range. It is so deeply ingrained in our ideas of the West that we are prone to overlook it, to forget its history and misunderstand its relevance to our present concerns. Its history, however, reveals the trade-offs and paradoxes that afflict western land-scapes economically, politically, and ecologically. Although the Western Range succeeded in obscuring the lines between public and private lands, it simultaneously committed the managers of those lands—both private ranchers and public agencies—to assumptions and practices that we now know to be flawed. Many of these assumptions continue to structure the debates among environmentalists, ranchers, and agencies concerning the use and management of some 485 million acres of rangelands in the West. In short, the Western Range was built on a foundation that can no longer sustain it.

The Western Range

On October 22, 1903, Teddy Roosevelt appointed a Public Lands Commission composed of three members: Gifford Pinchot, chief forester of the Department of Agriculture; Frederick Newell, the first commissioner of the Reclamation Service; and W. A. Richards, commissioner of the General Land Office. It was one of the most important acts in Roosevelt's career as America's first conservation-minded president. Pinchot wanted to clarify the status and use of the forest reserves, a new and growing system of areas withdrawn from the public domain to protect western watersheds from abusive mining, grazing, and deforestation. Newell likewise wanted to stop erosion, on and off the forest reserves, in order to protect the dams and reservoirs envisioned under the 1902 Reclamation Act. Richards was responsible for what remained of the public domain—all the land not taken up by homesteaders, prospectors, forest reserves, state land departments, or Indian reservations. For two years they conducted research and held public hearings throughout the West.

What they found was devastation. Livestock had flooded into the region during the cattle boom of the 1870s and 1880s, taking advantage of free and open land blanketed with grass. It was the paradigmatic case of the "tragedy of the commons": under the homesteading laws, settlers had been able to secure only small areas—less than a square mile—for private ownership. They had chosen sites endowed with water and fertile land, while everything in between had remained open to anyone who wished to use it. The result was disastrous overgrazing. Livestock had perished in huge numbers during blizzards on the Great Plains in the 1880s, and again during southwestern droughts in the 1890s. Cattle bones had been gathered into towering piles next to railheads for shipment to fertilizer plants. An early botanist in Arizona lamented that the destruction of grasslands there was so complete that he could scarcely find specimens to study. The winter rains of 1904–1905—which broke a six-year drought and remain to this day the heaviest ever recorded in the Southwest—blew out bridges, roads, railroad tracks, farm fields, and irrigation works from New Mexico to southern California. Earlier attempts to reform the laws affecting rangelands had foundered on political shoals, but after the turn of the century no one any longer disputed the need for change.

The mandate of the Public Lands Commission said nothing about ranching. Its assigned goal was consistent with the intent of the homesteading laws: "to recommend such changes [in land laws] as are needed to effect the largest practical disposition of the public lands to actual settlers who will build permanent homes upon them." Overgrazing and the erosion it caused were incompatible with making homes on the West's arid and semiarid lands, but by default, livestock grazing became the focus of the commissioners' recommendations. There was simply no alternative way for people to earn a living there. In a report submitted to Congress in February 1905, the commissioners wrote: "The great bulk of the vacant public lands throughout the West are unsuitable for cultivation under the present known conditions of agriculture, and so located that they can not be reclaimed by irrigation. *They are, and probably always must be, of chief value for grazing*" (emphasis added). The goal was homes, and grazing was the only means of achieving it—at the time, certainly, and as far as they could foresee into the future. Everything else about the Western Range followed from this prediction.

Free grass on the open range had been a major part of the profitability of livestock production during the cattle boom, but it became an equally significant threat to profitability once the frontier was closed. There was no longer any "virgin land" to move on to, and everyone familiar with the situation—ranchers, politicians, and the earliest range scientists—concluded that the open-access free-for-all was the source of the problem. "The holding or use of lands in common always results in rapine," wrote Jared G. Smith of the Department of Agriculture in 1899, "because of the principle that what is everyone's property is no one's, and no one is responsible for its abuse and spoliation." Describing the cattle boom, he wrote, "No thought was given to preserving the inheritance of those who were to occupy the land in future years; it was every man for himself, and he was the best man who could put the most cattle on the ranges to eat the most of the free grass." The solution seemed obvious: close the range by giving individual ranchers exclusive rights to graze their livestock on specific pieces of it. With no prospect of land-use change, ranchers would have every incentive to steward the range for its long-term health.

But how could this be accomplished? Turning over title to such large amounts of land was politically unacceptable—it smacked of creating a class of feudal lords—but leasing the land might achieve the same end. The commissioners studied the lease system Texas had adopted for its state lands in 1883, as well as one in Wyoming that dated to 1890, and they recommended a similar approach to the federal lands. Thus, in 1905, Congress reorganized the forest reserves into national forests, transferring them to the new Forest Service within the Department of Agriculture. Thus empowered, Pinchot quickly began dividing the forests into grazing allotments, which ranchers paid fees to use under ten-year, renewable permits. Closing the rest of the public domain remained too controversial to attempt until the 1930s, when the Dust Bowl and the Depression put the lie to homesteading as a viable settlement strategy once and for all. Finally, in 1934, the Taylor Grazing Act extended the Western Range model to the remaining public domain in the continental United States. In a last-ditch attempt to persuade Congress to turn the affected lands over to the Forest Service, the Department of Agriculture produced *The Western Range,* a ponderous report that documented the benefits achieved in the national forests since 1905. It failed in its main purpose, but succeeded in consolidating the view that exclusive leases were the solution to the problems of the range.

In theory, leases aligned private and public interests by creating stable, economic ranching units. Alone, the private lands were too small and disconnected to support a herd large enough to support a family.

Only in combination with public lands could private ranches survive, and the more secure the ranchers' leasehold was, the stronger the incentive for long-term stewardship would be. The theory was sound, but in practice, leases entrained a series of consequences that would ultimately undermine their effectiveness.

Dividing up the range was easy on paper, but on the ground it required fences. Without them no one could control the numbers and movements of livestock. Barbed wire was reasonably cheap, but the labor required to sink the posts and stretch the wire around tens of thousands of allotments was enormous. Larger, wealthier ranchers could afford it, but smaller ones could not, and the advent of fencing contributed to the consolidation of ranching into fewer larger operations over the first half of the twentieth century. Many ranch boundaries were not fenced until the Depression, when Civilian Conservation Corps crews were available to provide the labor. Interior fencing—to control the distribution of livestock within a ranch—was often not installed until further decades had passed.

Even more difficult than installing the fences was deciding where they should be built. Exactly how big should allotments be to support as many settlers as possible? If they were too small, ranches would fail; if they were too big, the number of families would be reduced and those that survived might resemble the feudal lords that leases were supposed to preclude. The number of livestock needed to support a family was relatively straightforward, but the amount of land required to support that number was something of a mystery. It depended, obviously, on how much grass the land could grow—but who could say? Different places grew different amounts; some had been more badly damaged than others, and how quickly they might recover was unknown; some produced fairly reliably, whereas others varied widely from year to year; some were flatter and easily grazed, others were broken by canyons and steep slopes that discouraged livestock from eating much of the grass. All these variables had to be incorporated into any robust determination of what a grazing allotment should contain within its fences. In typical Progressive Era fashion, the Forest Service turned to science to tackle this problem, summoning into being a new discipline.

Range Science

From the very beginning—even before the term "range science" had been coined—scientists studying western rangelands worked to determine the "carrying capacity" of the range. How many animals could be supported, reliably, on specified amounts of different kinds of land? Scientists arrived on the scene only after the damage of the cattle boom, so their job was doubly difficult. What had the carrying capacity once been? What was it now? Moreover, how did these two figures relate to each other? If one stocked a piece of range at its current carrying capacity, would it eventually recover to its "original" capacity? If so, how long would it take? To answer these questions required fencing their research plots and measuring the growth of grasses in the absence of livestock. This quickly produced powerful evidence to support the idea of fencing, a reform that was, by this point, virtually inevitable in any case. Photographs of fence lines dramatically testified to the resilience of the range once released from grazing, and some early scientists very nearly asserted that fences caused grass to grow. By clipping, drying and weighing grasses within their plots, then dividing the results by the dietary requirements of a cow, researchers soon produced the first estimates of carrying capacity, measured in acres per cow or cows per square mile.

The apparent objectivity of these calculations obscured as much as it revealed, however, and a few scientists seem to have recognized that they were jumping to conclusions. The numbers were accurate for the particular place and time they were collected, but could they be extrapolated to other places and times? Extraordinary efforts were made to classify and map western rangelands into types based on climate, soils, and vegetation, so that measurements made on a given site could be applied to larger areas of the same type. Even if this could be done, however, the problem of time would remain. What if forage production varied from year to year in any given rangeland type? One could calculate an average, of course, but how many years of data would it take to arrive at a meaningful norm?

These questions would not have been of much importance but for the economic and political exigencies of ranching and public lands management at the time. The Forest Service, ranchers, and the ranchers' financial backers needed carrying capacities that were constant and fixed. The Forest Service needed them so that it could allocate the range, enforce its grazing permits, and justify its decisions to politicians. The ranchers needed to know how many livestock they could run and what it would cost so they could plan their operations. And lenders needed to know what a grazing permit was worth on the market—what the net present value of a given lease was—so they could determine how much credit to extend to the ranchers. None of them had time or energy to expend contemplating the extraordinary ecological variability of arid and semiarid rangeland systems.

The result was no surprise: the scientists calculated carrying capacities as best they could; the Forest Service—and, later, the Bureau of Land Management—wrote the results into their grazing permits; and the bankers calculated accordingly. Whether ranchers stocked their allotments in keeping with permitted numbers is impossible to say for certain. The available documents suggest that they generally did, but very few documents exist for BLM lands, and stories of overstocking circulate like legends among ranchers and agency personnel alike. In view of what is now known about rangeland ecology, however, the issue of overstocking may be moot. Long-term measurements at the Jornada Experimental Range in southern New Mexico, for example, indicate that forage production is *less than half* of "average" *one year out of two.* In other words, variability is so extreme that even the permitted stocking rates would result in routine overgrazing during drought years. During prolonged or especially severe droughts, permitted numbers of livestock could damage the range in significant and lasting ways.

Conveniently, or perhaps tragically, these problems were effaced by ecological theory in tandem with conventional wisdom. The conventional wisdom of the time held that nature is intrinsically in balance—equilibrium, in scientific lingo—and that human activities could disrupt but not fundamentally rearrange this balance. Building on this view, Frederic Clements characterized vegetation as fixed by soil and climate for any given site. Human activities, such as livestock grazing, could alter this "climax"condition for a time, but nature would always push back toward its "original" conditions. If the human "disturbance" was removed, nature would invariably restore itself, healing whatever damage had been done. This theory emerged from the plains of Nebraska early in the twentieth century. Shortly thereafter, a Forest Service scientist named Arthur Sampson adapted Clements's theory to range management based on research he had conducted in the Wasatch Mountains of Utah. Sampson characterized grazing as pushing back against nature in a linear fashion in proportion to the number of livestock present in a given area. All that was needed to control which kinds and amounts of vegetation grew on a given piece of the range, therefore, was the correct stocking rate. Sampson left the Forest Service in the early 1920s for the University of California–Berkeley, where he became the nation's first professor of range science.

Clements himself did not advocate constant stocking rates for western rangelands. But his theory, as adapted by Sampson, became the scientific basis for the Western Range. It worked reasonably well in places like Nebraska and the Rocky Mountains, where there was more—and more reliable—moisture. In drier and more variable locations, however, such as the Southwest, the Great Basin, and on BLM lands in general, the Western Range system ensured chronic confrontation between agencies and ranchers. In a dry year, there would be too many livestock for the amount of forage, even if the rancher was stocking at the permitted levels; in a wet year, there would be much more forage than the permitted livestock could consume. In old Forest Service files, one finds letters and memos that document the back-and-forth complaints of the agency ("too many animals!") and ranchers ("too much grass!") through the middle of the twentieth century. Both were right, at least some of the time, and both were wrong in a more fundamental way. They were obsessed with livestock numbers when the most important factor driving vegetation dynamics was climate.

The New West
Ranches are turning into home sites in the New West because the premises underlying the Western Range are no longer valid. Many rangelands have degraded, becoming less productive for livestock. But even if the land had not changed, the price of land today would still exceed what livestock can justify from an economic point of view. The "chief value" of private rangelands today is not grazing but development. If

society's goal today were still simply to settle the West with permanent homes, as in 1905, then there would be no need to talk about ranching at all. Ranch owners, whatever their skills, values, or motivations, find themselves holding lands whose market value already stipulates land-use change, sooner or later. No wonder they defend their property rights so strenuously.

What we are watching—what ranchers are living through—is the messy playing out of this contradiction between an inherited set of rules, values, and practices, on the one hand, and a fundamentally changed economic landscape, on the other. Market forces are in the process of developing the half of the West that is private land, and some environmentalists are prepared to trade that for the end of grazing on the half that is publicly owned. They persist in believing that livestock numbers are what matter for conservation, and that the damage of the past century will heal itself "naturally" if livestock are reduced to zero.

The battles between ranchers and environmentalists over public lands have abetted the conversion of private lands, if only by distracting attention from it. Fifty percent of public lands ranches, according to one study, are supported by outside income or wealth; an estimated 107 million acres of private land are attached to public lands by grazing permits. If one were to measure these private subsidies and add them to the opportunity costs of not subdividing, one might well find that ranchers subsidize ranching to a much larger extent than do federal and state grazing and tax policies. We have learned that the Western Range demanded too much uniformity from the land. Spatially, it sought a single framework for the entire West, to match the scale of federal land management agencies. Temporally, it sought stable stocking rates for each allotment, each ranch, year after year. Yet the contention and litigation of the "rangeland conflict" have only made agencies less flexible and adaptive, further inhibiting effective management.

The Western Range is now like a leaking lifeboat: a major problem, but one we cannot simply abandon. By linking public and private lands together, economically and administratively, it has helped prevent the conversion of hundreds of millions of acres of land to more intensive uses. It has kept the West less fragmented and closer to its native vegetation than any other part of the continental United States. But to maintain these legacies, and to sustain ranching rather than simply settlement, we have to develop an alternative set of rules, values, and practices, and somehow graft them onto the Western Range or incrementally put them in its place. We have to rebuild the ship while continuing to sail in it. This process has already begun at countless locations across the West: land trusts buying conservation easements on private ranchlands; groups of ranchers finding alternative markets for organic, predator-friendly, or "natural" beef; nonprofit groups holding rangeland—both private and public—for the use of neighboring ranchers during drought or while conservation actions such as prescribed fire are implemented. These efforts seem always to be local, grassroots affairs concerned with a watershed, valley, or similar-sized landscape. Reform of the Western Range as a whole is hostage to federal-level political stalemates, and it is unlikely that any single solution could work across such a diverse region. Only practices tailored to particular circumstances of ecology, economics, politics, and personalities can possibly succeed. Perhaps the most difficult questions are those of scale: Can ranching survive on isolated patches of land—particular watersheds or valleys, the scale at which local groups can craft new social conditions for it—while the surrounding landscape fragments and develops? Is western ranching dependent on a single lifeboat, or are there many, some of which may sink without endangering the others?

Along the San Pedro River, the contrast between the city of Tucson's two forays into ranch ownership demonstrates the difficulty of these questions. As agriculture has stagnated since 1960, subdivision and conservation have emerged as near-equal rivals in the struggle for the middle reach of the San Pedro. More than 11,000 acres of private land has been subdivided, and the fate of each parcel that comes on the market is closely watched in the community. But the road is still not paved, and developers who've tried to launch big subdivision projects have all failed. Houses have been built on fewer than half of nonagricultural private parcels, whose average size is a whopping sixty-eight acres. As private land, these parcels are more protected from certain threats—recreation, for example—than they would be if they were publicly owned, and their absentee owners largely leave them be for the birds and other wildlife that frequent the area. Meanwhile, public agencies and private groups (such as The Nature Conservancy, which declared the San Pedro one of the world's "Last Great Places" for biodiversity preservation) have bought properties and grazing leases totaling more than 100,000 acres. Whether these lands can help keep local livestock operations viable—by providing emergency pastures during drought, for instance—remains to be seen. As long as the road remains unpaved, local people may be able to bail out the Western Range faster than the New West can flood it with outside capital.

Teresa Jordan was raised on a ranch in southeastern Wyoming and is the author or editor of seven books. Of this poem she says, "I've always been incredibly grateful that I had the chance to grow up on a ranch. I loved the freedom of being able to move in the world under my own power. In college, when I lived in an apartment complex with many young families, I wondered about the children who played on the swings in the courtyard. Did they ever yearn to be alone? Did they have secret places of their own? Once you have experienced such things, they are always a part of you."

Looking Back

Teresa Jordan

The secret place is gone.
Picked up like a tenant
in the middle of the night
after a bad run of luck
it trudges down the dark lone road
with the meadow
and the barn
and a long line of cows,
tails bedraggling behind them.
I loved

that secret place
down by the riverbed
hidden by a bank. I whittled
dolls from willows there, made whistles
out of broad bladed grass, told my big bay
Buddy how I'd never leave.
I lied

though not from will.
Let me be salt
sculpted by cow
tongues until I am lace
and then I am gone.

I want to belong to the ground
again. It is the barn

that breaks my heart
trudging soddenly along, bedsteads
and broken harnesses rocking
softly in the loft, lost
beneath great drifts of
guano. A spavined horse-
collar mirror hangs
cockeyed on the ladder
and that other me looks back
amazed. In the darkness
only one of us is
gone.

Laura Pritchett is the author of two works of fiction, both set on ranches: a novel, Sky Bridge, *and a short story collection,* Hell's Bottom, Colorado, *which won the PEN USA award and the Milkweed National Fiction Prize. In this essay, she writes about her family's ranch in northern Colorado, her eight siblings, her parents and their herd of cattle, and of the struggle to say good-bye.*

Hoof Making Contact

Laura Pritchett

I have come to believe that both my physical life and my spirit are so deeply connected to that particular plot of land, the family ranch, that I might be a stalk of grass myself, rooted in arid and meager soil.

—Linda Hasselstrom, *Feels Like Far*

The truck barrels down the rutted, snow-packed road that stretches across our ranch, and I'm going too fast but don't care because there's a good song blaring on the radio that has me tapping my fingers and singing off-tune and I'm feeling light-headed and free since I am alone and outside on this February morning. I lean on the horn and, as if in response, the pickup sends me flying up as it jolts through a deep puddle pit, and I honk some more but this time I slow down, agreeably adjusting to the truck's request. In between blasts of noise, I lean my head out the window into the cold air and yell, *Come boss, comebaws, c'mbaawws!* which I do as some sort of joke with myself: the cows don't need to be called, it sounds ridiculous, but everyone does it anyway and here I am, too, mucking up the world with unnecessary words and noise. Exactly the thing I would normally be loath to do—coming out here with the radio on, horn honking, voice propelled into the quiet world, but today it seems all right to have a little fun because well, what the hell, and also, if I don't, there's a chance I might sink into a quiet sadness where I do not want to go.

The ruddy-red cows are coming as fast as they can, having broken into a lumbering trot at the first sight of the truck. They are hungry, not only because the pastures are covered in snow, but because of the calves inside them. Curled up in the final stages of forming, the babies make the cows' bellies bulge, their stomachs hungry, and the herd looks wonderfully ridiculous coming at me with stomachs swaying to-and-fro, and I urge them on with a *Come-on-you-sweet-mamas, come on now.*

I park the truck in the middle of the back pasture, take it out of gear, and climb in the bed of the pickup with the bales I've brought out this morning. I brace my knee against a bale and pull, bending the rectangle of dried grass enough so that I can slip off the orange twine. I throw flakes of hay down into the snow, leafy green rows on either side of the truck. Some of the bales are moldy, and a fine gray dust flies out at me as the bale separates. I turn my head and push my nose into the flannel of my shirt until most of the cloud has passed and wisped into the air, and when I look up I notice: now the world is quiet. It has settled around me like fog, a stillness quivering with the smallest of intrusions: a killdeer's song, the noise of the river, the cows' hooves crunching snow. It's a different kind of light-headed happiness I feel now, a quieter and truer sort.

After I pull the truck forward a little, I throw out more hay. I do this several times until two lopsided green lines of hay cross the white pasture. *Come boss, c'mbaawwws.* This time my voice is only above a whisper and anyway, most of the cows are already here, pushing each other aside and scrambling for the food, closing in on each other to block out the few stragglers coming up from behind. The cows are earnest in their eating, but some of the younger ones are playful, kicking at each other good-naturedly, pushing heads, snorting and shaking their noses, and I scratch one such cow on her head and offer a *Yeah, that's how I feel today, too.*

Though I moved off this small ranch years ago and am here only to feed in my parents' absence, I still recognize many in the herd: Elf Ears, who had her ears stunted by frost as a calf; Big Mama, who always looks pregnant even when she's not; Old Mangy, who looks rough and tired; and a younger one, whom I know only by number, who has a playful, teasing light in her eyes and I decide to call her Happy Day, because something about the pulsing blue sky and snow makes me think that such a day works pretty damn hard at pushing fears and sadnesses away, and that at some point the creatures of earth just ought to give in and listen.

At the last stop, I stand in the bed of the truck, jabbing my fingers in the air, counting. Then I hop off the truck and mosey around behind them, looking at their rear ends. Calving season is still weeks away, but I want to make sure I don't see any signs of labor: full udders and teats, tails to the side, calves dropped into birthing position. Perhaps it is their attitude that tells me the most: they are not yet sullen, quieted by an oncoming birth. Instead they are rowdy, shuffling and nudging each other, even kicking at each other with their back legs.

All looks well, so I take a moment to look across the ranch toward the first foothill, a rocky ridge spotted with yucca and mountain mahogany sticking out from the cover of snow. Atop the highest rock sits a blue heron, facing the river. I follow its gaze and see a hawk circling. The horses are bunched together in another pasture, and if I squint, I can see the flares of mist escaping their noses. To the east are the white farmhouse and barn, bullpen, corrals, chickenhouse, sheds, all clustered together around bare cottonwoods. Up on the hill rests the old cemetery where some of the town's early settlers are buried. I turn around and around, like a child's game, except I do it in slow motion. In this way, I can see the stretch of land below the foothills to the Rocky Mountains. As a child, I believed this to be all mine, but now I know that I ought to start saying good-bye.

As I get in the truck, I see that one of the cows, in nuzzling the leftover hay in the bed of the truck, has pulled an orange loop of twine down. I don't want to leave it there, since the string can bunch up in a cow's intestines and wreak havoc. I bend over to pick it up. A crow squawks in the far-off distance, I wish for a cup of coffee, I hope to hold on to this light feeling for a while. Then I feel it: the knowledge of a danger, the presence of an oncoming hoof. I exhale. I don't move—I don't even know which way to move—and then it is there, moving right past my left ear, scraping my skin only lightly.

I laugh—out of surprise, or perhaps because the moment was nothing after all though it felt like it was. I stand up, take a step back and look at the cow that has just kicked at me. She is calmly chewing the hay crammed in her mouth and pauses to rub her head against the stomach of another cow.

A fraction of a second, that's all. But this moment will become slowed down in my mind and I will later think of it as a close call with serious danger. The speed and momentum of the hoof, the way it would have cracked against my skull; these give me pause in a certain very slow, very still way. Often I will remember this moment as a strange mix of light joy and sudden terror and how both held on with equal strength in my heart. And often, I will think of how lucky I was to just hold still, to pause, to let the danger come near but not catch me after all.

It's killing us, my mother e-mails her nine children. *Something has got to be done with this place. None of you kids love us much, or you'd notice that, and I repeat, this place is killing your father and me.*

She describes possibilities for the land's future, interjects her own opinion, tempts us with figures. The family would get a few million if they sell the place to developers; more if they let the place "be graveled" and store water for the city in the huge pits of earth that would be left. They could put a conservation easement on all or part of the place. They could let the kids build small houses on the eastern side, the side already bordering development along Overland Trail, and let the rest remain open space. They could sell it to a rancher or wealthy landowner who would keep the place intact. They could donate it to the university.

I stop in for a visit and am warned by my mother: She's putting a For Sale sign out along the road tomorrow. She's turning it into a bed-and-breakfast, only no—then there'd be sheets to wash. She's going to give it to some nuns. She's going to give it to the next person who drives in.

Now, now, says my father.

You kids aren't worth a dang anyway, my mother adds.

Well, now, says my father. *That's not fair.*

She bows her head and turns away from us so she can go inside and cry alone. She is that tired: of sifting through options, considering taxes, thinking of children, not getting any help.

They both want to be fair. They want to keep everyone happy. They want to do what's right. And mostly, they want out.

The place is a lot of work—constant physical labor and the inside-at-the-desk variety as well. It's a small ranch, to be sure, but it's more than enough for two aging people: cows must be fed, checked, doctored, birthed, weaned. Endless weeds need to be killed, fields irrigated from May to October, trees watered, fences mended. The garden needs tending and the apple trees pruned. The pump needs to be fixed, several sheds are in disrepair, the cows escape regularly. The peacocks need to be fed, the chickens shut in at night, the horses ridden, irrigation pipes monitored in the summer, indoor plumbing kept from freezing in the winter. There are meetings to attend, taxes to pay, books to balance, cattle to sell.

All this for nothing, in the common understanding of things, meaning that in all the years my parents have owned this ranch, no profit was ever made, and in fact, all this effort has been exchanged for a loss of thousands of dollars per year. The place proved to be a great tax write-off and investment, but created a cash-poor existence, and this fact translated itself into duct-tape patches on kids' shoes and the heat being kept very low and strict limitations on water use and food consumption and other stories of living below the poverty line that we kids share now, as adults, sometimes still with the pang of embarrassment or hurt that such poverty brought, this poverty which became, thankfully, less overbearing as the older ones moved out of the house and the younger entered teenage years. But this land was a place to belong to and a place to raise children, a hobby ranch that was so small it could never be economically viable and it didn't matter, because, as is often the case, hobbies are also loves. Since my father worked for income elsewhere, as a professor of animal genetics, there was the knowledge we children could be supported, if humbly, and so this land was chosen as the recipient of all extra monies and efforts, and for that reason some grew to hate it as it meant sacrifice and others grew to love it because of that sacrifice, and in any case we grew up riding horses and chasing cattle and fixing fence and bringing newborn calves into our kitchen in the middle of the night. We grew up and left, and what remains are layers and layers of work for two people who no longer can do it all.

So sometimes, my mother calls me: Dad is out feeding again, he's working himself to death, and if you love him, you'll get out here and help.

I love him, I love him! I'll drive out to help, I'll be right there! But so many chores can only be done by a person actually present—if a fence breaks, if irrigation water comes a little early, if cows get out, if a horse founders—for all these things, a person must be physically there to help. And there's no place for my own family—my husband and two young children—to live there. So we live elsewhere, only a few miles away, but far enough that I'm not there to see and respond. I am busy writing, changing diapers, making dinner, washing dishes, hugging a sick child, doing laundry, tending a garden, editing a story, kissing an infant on her nose. I want to do all this, and I want to help my parents, and I want to protect the ranch, and so when my mother calls I bow my head and with my palms, press the tears back into my eyes.

In the meantime, we kids are presented with options. Some voice opinions loudly because they fear they otherwise won't be heard, and others refuse to comment at all because they doubt their opinions matter. *Piss on it,* says Edmund, *nobody ever cared what I thought anyway. I don't know what to do,* says Al softly. *I'd like to live out here, but it'll never happen,* says Charles. *I just want a teepee,* says Bernadette. This what-to-do-with-the-ranch conversation reaffirms understood alliances and different perceptions of the past: there is the camp that has had enough of this family, the camp that still coheres strongly, the camp that has no interest or time to care at all.

Of those who speak, most are careful not to be pushy or greedy or demanding—to turn into the circling vultures our mother warns us we might be. Instead we remain what she also accuses us of: being so warily watchful that inaction is the only result. We mumble things about it being *their* decision, and what do *they* want, and besides, we acknowledge that every option is so complex—taxes and legal agreements and rules and regulations and documents—that after a while, most sink into the it's-too-complicated-you-decide routine. All I know for sure, now, is that I see a hoof recklessly swinging in this direction. Whether it makes contact or not remains to be seen.

Turn it into a golf course, e-mails my brother Raymond from his home in Boston. Luckily, this is one option the other eight of us can discard easily with a laugh and bit of eye-rolling. The last thing this arid-climate landscape needs is another water-needy golf course, and anyway, Raymond's ideas always seem a bit suspect, only because he's been on the East Coast for so many years, and returns so infrequently, that he can't help but be slightly out of touch.

Brother Jake wants the money. As much as he can get. He's all for graveling the place. He admits that for five or so years, the land will be wrecked-up and ugly. Bulldozers, semi trucks, deep pits in the earth. But look, he says, the world needs gravel. And local cities need a place to store water. And in the end, if landscaped correctly, what will remain is an attractive fifty- or seventy-acre lake. With proper care, it could provide great habitat for ducks and wildlife. Certain funds could be set aside to restore it—bushes and trees and natural grasses introduced and tended to. This option could be combined with later family development. Because of the graveling, we'd all have the money to build the log cabins several of us yearn for, or set up teepees as Bernadette suggests, or to start a vet clinic there as Charles desires. Jake also points out: most of his lame siblings still live near or slightly above poverty. Several million dollars would be a lot. Edmund could get help for his autistic son. Al could get braces for his ten-year-old. Kary could quit her minimum-wage job. Charles could quit ruining his back shoeing horses. *You are all stupider than I thought*, writes Jake in an e-mail, *if you're going to turn down this money and p.s., no, I am not being a selfish jerk, I am telling you to watch out for yourselves and your children.* I write back that I don't need any money. He responds by saying I haven't been kicked in the ass enough by life to understand hardship, to understand what giving up this money would mean, and I write back and suggest that it's possible I have, and he responds only by saying I must then be a very slow learner.

Kevin, Charles, Al, and sometimes I, depending on the day, would all like to build on the place. We would cluster our houses and leave the rest as open space—but would we run it as a small cattle ranch or something else—could we manage it together? The county rules on this are complicated; various maps are drawn up, meetings are held with the county commissioners, but clear answers are few. And what, really, would be the logistics of this? Can we really live next to each other? Do we really want to? Mom reminds us that none of us could afford to build more than a basement on the place, let alone pay for utility lines and roads. She also keeps reminding us that in her opinion, the greatest gift a parent can give is freedom: to nudge kids out of the nest and then yank the nest away. *I don't want to leave you anchored*, she says. *That would be the worst thing I could do.*

Well, the anchor is there.

I'd be lying if I said I didn't want to live there. I do, since there is no other way I could ever get a patch of land outside town and build a log cabin and have a donkey and honeybees, and a chicken or two, which, plain-and-simple, is my *dream*, a dream otherwise unattainable because land prices are so high and open spaces so few—especially in a place like this with the aesthetics I call my Soul Place, for lack of a better phrase, meaning that it feels exactly right, and for me that includes either the anonymous place in the Arizona desert I passed one time on a car ride years ago and was struck with the utter *rightness* of that place, or Walden, Colorado, a place that takes my breath away every time I pass through it, which is a number of times per year, or a place much like my family's ranch with irrigated pasture and beachy paths next to a river. Places that seem right like this are few and my heart cries out for them to an extent even I find to be silly and sappy but nonetheless I feel the way I feel. So yes, I want to live there.

I ask that the entire place be put into a conservation easement. To leave the place alone, unchanged, forever. I e-mail everyone a little rhapsody about why the land should remain as is, why once it's paved over or dug up, something of its soul will be lost, why its preservation is paramount. Don't they remember walking down the gravel lane with fishing poles in hand? Don't they remember sliding across the snow in a tractor innertube being pulled by the truck? Don't they remember galloping a horse up the back hill, finding arrowheads, floating down the river and yes, fixing fence, collecting eggs, birthing calves? How can they imagine such a place *paved*? Dug up? Trampled, shifted, ruined? Leave it alone. Let's leave it alone. Let's give it the gift of our absence.

I get sympathetic responses from one or two siblings. Some want to see a conservation easement combined with limited development. From three or four, I get e-mails filled with impatience and condescension for continuing to be so naive and idealistic well into adulthood. Where do I get my high-and-mighty ideology and can I please keep it to myself and along the way, try to grow up? A conservation easement will reduce the land's value significantly, tie the family's hands, put siblings' futures in jeopardy. I should get some perspective: people have been using this land for centuries—using it to best fit their needs and only bounded by their technologies. Besides, the whole area is "ruined" anyway. Development is so rampant in the area that it's not like this small parcel is going to save the wolf, the mountain lion, the bear (though the latter two are taking up residence on the place right now). What do I envision I'm saving, anyway? This is no paradise; it's a tiny piece of land crowded on all sides by people, who need a place to live too. Also, who will manage it? It's simplistic to think it could go back to native grasses. It won't. Weeds will take over the place; without my father's battle to combat thistle and leafy spurge, the grassy fields will be overrun. (*Get goats and sheep?* I suggest, to which no one responds.) Anyway, my parents have discussed such an option with several groups: it's too small to be of interest to The Nature Conservancy and so far, no other group has offered to manage it. Interested groups want my parents to remain on the place and continue to do the work, and this is something my parents cannot do. The county has offered to buy a partial easement, but wants to put a bike trail down the middle, something that my parents strongly object to, and the city has offered to buy it as well, but similar problems would ensue. But yes, a conservation easement is something they're considering and I smile, catch my breath, hold still.

For a while, advice and opinions get thrown back and forth across the Web. Someone suggests that we talk through the realities of living next to each other. This prompts us to start talking behind each other's backs, or, as is more often the case, *not* behind each other's backs: Someone's house will be junked up. Someone else would never do his share of the work. Some people don't like other people's spouses. There are real differences in ethics. Tangential issues get brought up: so-and-so's depression, temper, finances, family, and my—and here I brace myself for an attack on my liberal politics, my I'm-sure-I'm-right attitude. Instead I get a few stings about, of all things, my good-natured disposition, which is found to be genuinely offensive to some, who imply that this attitude can only be the result of deficient intellect or childhood amnesia or both. The rampant e-mail squabble goes on for some time.

For the first time in a long time, I seriously consider my family. I realize that several of us don't keep in contact with others, not even at Christmas or birthdays—the only exception being the flurry of e-mails about the ranch. *Why?* I wonder. Perhaps there are too many of us, gone off in too many directions, too busy with our own lives. Or maybe it has to do with the fact that several left this place in order to leave their past behind. Perhaps it is our personalities. Most are introverts, but some are uncomfortable around other people in the extreme. I realize that only one of nine currently has a job that involves working for someone else. There is an independent stubbornness and a hardness. There are very deep and sincere bonds, to be sure—but all nine of us getting along? I was the only one who believed it could happen, and I got enough you've-got-to-be-kidding responses to give up in defeat.

This discussion about what to do with the ranch, prompted as it is by the knowledge of my parents' increasing age and declining health, has started tentative communication where none was before. But soon the e-mails lapse; everyone is too tired of everyone else to think anymore. Another year goes by. My mother sometimes writes us: *This place is killing your father and me.*

Another winter is coming. This past weekend, at the end of October, I am asked to come help round up cattle—the calves need to be weaned and the cows' pregnancy checked. My father drives the pickup ahead of

the herd honking and calling *Comeboss, c'mbawws* and most of the cattle follow obediently. I walk behind the cows, keeping them going in the
right direction, chasing back the belligerent few who try to turn around.
It seems that all the whips have disappeared, so I swing an old bamboo
ski pole about good-naturedly, whacking a cow or calf on the butt only
when they turn around to seriously challenge me.

Once we are in the corrals there is the ruckus of separating mother
from calf, but after enough dodging-around hand-waving cussing-work
they stand in different pens, bawling at each other. Then the cows are
run through the chute and I am in charge of filling syringes and keeping the records—the job that has been mine since I was a little girl.
Miraculously, all the cows are pregnant, five months, and they'll give
birth at the end of February or beginning of March, and there is the
usual banter about them all going into labor the same night, that bull
doing a good job, whether or not number 701 should be sold because
she's got bad feet, the bad corral system, the cow shitting on the hand
of the man who is reaching inside her to feel her unborn baby.

The calves keep bawling and the cows bellow back, and this noise
will continue for the next three days as they are kept separated. The
cows' milk needs a chance to dry up so that their energy goes to the
making of their new calves instead of nursing their yearlings, and for
once I don't feel sorry for the calves because they should have been
weaned long ago and their mamas are probably tired from nursing all
this time, and yes, a time comes when a baby needs to be pushed away.

At the end of the day, as we walk out of the corral together, I refrain
from turning to my father. I don't want him to see me protesting my
separation from him or this land that is so much a part of him that
sometimes the two are equated in my mind. I know the look I have in
my eye—I have seen it in my own toddler's eyes and I'm sure it has
seeped back into mine—a look that demands, wants, begs. Instead I
keep my eyes on the land in front of me.

Lately they have been talking of selling the place to a developer for
high-end, huge houses. McMansions, I call them. We both know I object and that I object strongly. We both know that it is not such a remarkable thing to forgo money and the relief it can bring, it is not
impossible to set aside a dream in order to realize another, but also, as

I now understand, it is not impossible to give up in defeat and step back from a decision and a place that is not mine. This past year I have been saying good-bye, and now I am nearly ready.

For this moment at least, I allow myself to entertain dreams, and as my father and I walk to the house I list to myself the things I want. To protect this ranch, to help my parents, yes. And to stop the ache in my heart that comes from desiring something too badly. Because somewhere, someday, I want to drive out back on a cold winter day and bring hay to hungry cows, sometimes with music and light making me giddy and sometimes keeping myself and the world quiet and still. I want to be a steward of the land, a very good one, and teach my children to do the same. When I take them arrowhead hunting or fishing or cross-country skiing, I want them to understand themselves within the parameters of the natural world, and I want that for me, too. I want them to know the force of a bond to a beautiful place, even though I know that, as with all new loves, some pain is apt to come flying at them, and will surely make contact.

The best I can do, perhaps, is to hold on to the light joy of a cold morning—and hold steady when danger comes and hold still and hold still and *feel* it.

Kim Stafford departed from graduate school to be with children in Wallowa County, Oregon's outback, in 1978. "A Buddhist in Cattle Country" speaks of affection for the kind of unusual encounters that happen far from the odd certainties of city life.

A Buddhist in Cattle Country

Kim Stafford

We heard a rumor she was on her way
to spend time in a quiet place alone
and met her first at the barbeque
out to the ∠ Ranch where

you had to see it through
her eyes: smoking carcasses
on spits turned slowly
over the mesquite coals.

God, that meat was good.
Burl knows how to fix it right.
You had to give her credit,
too: she sipped a beer

and asked how cold it got
hereabouts, and where
could you go for books,
and was the sky always

amazing and gold like that?

Joan Chevalier is a speechwriter based in New York, where she works primarily on Wall Street. A Blue State liberal with guts and a mission, she ventures into a Red State stronghold of cowboy poets, notorious dogs, rodeo queens, and altitude sickness to give us "Wasted on a White-Collar Job." To satisfy her curiosity about the West, its politics, and its people, she participates in a Wyoming roundup, where she looks to find a balance, an explanation, a radical center.

Wasted on a White-Collar Job

Joan Chevalier

Our country is fast approaching a precipice; we are about to make decisions about water, trade, and land that will be ruinous to both rural communities and the landscapes we tend; decisions that I am afraid our nation will long regret.

—Pat O'Toole, Wyoming rancher

anchers never tell you what to expect. So I received instruction from a fellow ranch guest and good friend, Rick Knight. Rugged and capable, Rick might pass for a rancher, except for this: unlike ranchers who keep a tight rein on speech, Rick is almost unnervingly garrulous. A wildlife biologist and professor, his conversation is mostly instructive and always characterized by the kind attentiveness of all dedicated teachers. He pointed to the approaching pregnancy tester and told me, "You can tell he's a lefty from the brown stains on that side of his coveralls and hat." I recognized that he was trying to mediate this New Yorker's first experience of a roundup. It didn't work. I would have to ask the nature of the brown stains, about which I had my suspicions, and there simply was no room left in my coping for that answer.

As though undaunted, I took my place on the line at the corral. The cows were herded one by one down a narrow maze into the chute, where they were secured for the pregnancy tester. Casually plunging his plastic-wrapped left arm to shoulder depth up one cow after another's rear end, an endless, seemingly self-replicating line of cow rear ends, the tester pronounced his verdicts: "open" meant barren and death for the cow; "okay" meant the cow was reprieved by her fruitfulness and would return to another pasturing cycle, moving from high-mountain aspen groves to sagebrush flats to the irrigated meadows of the home ranch. That much I pieced together. What I didn't see coming was the dollop of cow shit that landed in my eye, whipped there by the irritated flick of a mother's tail. From 7:30 a.m. till 4:30 p.m., not stopping to eat or drink or pee, splattered with cow shit and snot, there I stood: closer to a pregnancy test than I ever intended to come, well and truly content in the just discovered assurance of my own capacity, my own—who would have guessed—hardiness.

Carol and Richard Hamilton's ranch lies on the Wyoming side of the Uinta Mountains, down by Fort Bridger in the state's southwest corner. Theirs is not a "dude" ranch for tourists, but a working family operation. I flew in from New York City for the roundup two days earlier than planned. A premature snow in the mountains was driving the cattle down and out of compliance with the National Forest Service timetable, a program that does not accommodate acts of God. I knew the mountains from my first trip three years ago and Carol's cautionary instructions: "always drive toward the mountains if you want to get to the ranch." The distant Uintas, into which the ranch extends by virtue of national forest grazing allotments and checkerboard-patterned private holdings, were far less imposing in the July of that first trip. Now, in late September, the snow-mantled peaks glared and menaced in the sun, remote and massive, a mind-numbing rejection of my merely human self. We began the roundup there.

Not quite up into the snow, not above the tree line but within its chill breath, the first day's ride required locating well more than a thousand head of cattle, owned by three ranch families, and bringing them down from the mountains to the sagebrush flats. The home range lies at 7,000 feet above sea level; the first day in the mountains, we rode at 8,500 feet. We began at 5:30 a.m. when the temperature still hadn't climbed above freezing. We rode without saddlebags, without water or provisions. We rode till the work was done. We did the same the second day, pushing the cattle across the sagebrush flats back down toward their respective home ranges.

The first day's work in the mountains was difficult by virtue of the terrain: designated wilderness allotments without trails, but with dense brush, low branches, bogs, streams, precipices, and the threat of worse horse-spooking stuff, such as coyotes, bears, and camouflaged bow hunters. The second day's work on the flats was even more intimidating because there the cattle were gathered in full force as a handful of riders orchestrated pushing miles of widely dispersed cows in one direction, sometimes a direction that a section of the herd, or just one mother and calf, decides is the wrong one. And therein lies the nightmare scenario: the entire herd becomes convinced that the rancher's compass is in error and their fellow cow has the better idea.

At a time when no other better riders were near (which would be all of the other riders, including the children), I was unfortunate enough to have a calf decide that his mother must be back up in the mountains. Mewling belligerently with a baleful, stubborn look at me and my horse Chester, the calf darted off on his lonesome. I figured that, if old Chester and I didn't get him turned around, he was a goner. So, witlessly, with no sense of my own severe limitations as a rider, I urged the ancient horse on with "Get him, Chester." I had no idea what I was summoning up from the old brown horse, who suffers from several age-related infirmities. I had no idea *any* horse could be so swift and agile. We tried to corner that calf and turn him around in the right direction five times. Finally, as I barely hung on through Chester's swift, sharp maneuvers, it dawned on me that I was not the rider for this endeavor. I reined Chester in, just a tad, to bring him from a full-out gallop down to maybe a full-out trot. But that was enough of a signal for Chester, who then agreed that I was not the rider for this task.

Above all else, Chester has his reputation to protect: he has never dumped a rider, no matter how inexperienced and no matter how dire the circumstances. Even Richard Hamilton, who resists all sentimentality, acknowledges that Chester is one of nature's great boons, a bombproof horse. Richard tells the story of the Japanese student in range studies whom he and Carol came to love (they named a dog after her), but who was an abysmal rider. "There she sat on Chester, straight as a board, the reins up to her neck, just like Hirohito. She didn't look left or right; she just stared at the back end of the cow straight ahead. Then, Chester got going. He swept left; then he swept right. He might as well have had a porcelain teacup on his back, but he moved that section of herd out all on his own. And I thought to myself: *There's* the testament to how great that horse is." As ranch stories are not meant to be lost on the listener, I remembered this and heeded Chester's inclination. We left the wayward calf to his own devices. I learned later that, in all probability, he would be rounded up when they combed through the mountains again in a few days, that you expend some energy on the strays, but not all of it. As was to be expected, Chester made the right decision.

On the third day, the riding was lighter, less than a handful of early dawn hours before we began the corral work, where I had to contend only with flying shit, snot, and hooves. Blessed relief, as my saddle-bound knees were now in constant conversation with me: clicking and clacking where I prefer my body parts silent. Master that I am of the self-pampered life, processing hundreds of cows at the end of three days of hard riding was the culmination of an odd passage for this liberal, green, horse-fearing New Yorker. In the summer of 2002, uprooted from my place in the world, an office at the World Financial Center, and on nothing much more than a post-apocalypse whim, I spent my vacation in a rangeland management class at the University of Wyoming. The class required I live for a bit on a working ranch, Carol and Richard Hamilton's ranch. My world was changing more than I imagined possible.

Not many months before the 2004 presidential election, the Wyoming Stock Growers' Association wrestled with a proposed vote of "no confidence" in then secretary of agriculture Anne Venneman. Unrest verg-

ing on revolution in the presumed Republican rank and file. Throughout rural America, there was, and continues to be, a growing outcry
with regard to trade and agricultural policies that disadvantage the family producer (Karl Rove's intransigence with regard to country-of-origin
labeling, for instance, which he continues to insist is "trade protection").
And still, John Kerry couldn't capitalize on it. As one "progressive"
rancher, an unrepentant Democrat, angrily told me, "It's this simple:
no democratic presidential candidate can win here now, not after
[President Clinton's secretary of the interior Bruce] Babbitt's War on
the West."

There's a snap and snarl these days to the traditional crankiness of
the rural West. In the words of a political scientist, William Chaloupka,
crankiness has turned to downright "churlishness" and the "target is
most often green." This largely green backlash transformed the interior
West into the "most thoroughly one-party region the country has ever
seen [and] ... any number of green issues foundered." His essay on the
environmental movement and the alienation of the West is one of several in *The Politics of Moralizing*, described as "illuminating how to fashion affirmative discourses and practices that resist the self-righteousness,
resentment, and deafness of moralizing postures across the political spectrum." But apparently, the left now thinks those postures belong only
to the right.

What amazed me most in the flummoxed East Coast analysis of the
2004 presidential election was what was absent: curiosity about the
largely rural regions that gave President Bush the election, curiosity
about the farmers, ranchers, loggers, and miners who didn't see it our
way. As a study commissioned by the W.K. Kellogg Foundation concluded, the Republican Party's "standing as a 'majority party' depends
greatly upon their strength in rural communities." Even my home state,
Pennsylvania, went to Kerry only by virtue of its urban voters. But in
the hardened-to-a-prejudice view of the coasts, the heartland is a red
monolith of empty space and empty heads; the great rural reach of the
country mired, according to the postelection fulminations of Blue State
writers, in "ignorance."

At one point in my concern for America's rural communities, I met with Undersecretary of Natural Resources Mark Rey at the USDA, to discuss the Tongass National Forest and its all but lost logging community. Somehow, in the midst of that conversation, ranchers came up, and I did what I always do now—I cried. I was grateful for small favors: that those were silent, streaming tears as opposed to my more usual sobbing, choking, bedraggled deluge. Nonetheless, there was Mark, whom the environmental community calls "Darth Vader," reduced to pillaging his office for baubles in an attempt to distract me: "Look, here's an interesting piece of wood" or "What do you think of this book?" or "How about a cup of coffee?" Then finally he said: "A rangeland management course at the University of Wyoming? Talk about jumping into the deep end of the pool." Thus reminding me of the gumption it took to take the course (my small portion of gumption), and my tears were staunched. In the end, I may have lost my dignity, but I gained a great deal of Mark's personal office booty. He stopped short of giving me the stapler—that would have been government property.

Mary Flitner, another Wyoming rancher, was the instrument of my delivery to that rangeland management course. Mary, like Carol and almost every ranch woman I know, is small and wiry with a handshake that could bring you to your knees. I have come to understand Mary's "suggestions" for what they are: moral challenges proffered with a large measure of kindness. It is the kindness, I imagine, of the Hindu god Ganesh—my favorite for the sole reason that he is an elephant—smiling as he says, "You don't *have* to do this, you know; transcendence is not mandatory." But Mary has such faith in the transformative power of direct experience, albeit difficult direct experience (not martinis on the Riviera, my preferred kind), that I couldn't disappoint her. Well, actually, I just couldn't come up with a good enough excuse.

I met Mary and Stan Flitner in Manhattan on September 19, 2000, at a roundtable discussion on conservation and public lands management convened by the Theodore Roosevelt Association and the Franklin and Eleanor Roosevelt Institute at Theodore Roosevelt's birthplace. A decorous, staid brownstone on Twenty-eighth Street, it is an unlikely home for the vital, larger-than-life TR and—except for it being his birthplace—an unlikely setting for me to discover American ranching.

The usual (or maybe not so usual) suspects were in attendance: the heads of all the lands agencies, the heads of some environmental organizations, a handful of "outdoors" writers (not including myself, as I am generally an indoors writer), a handful of wealthy concerned New Yorkers, and then, Stan and Mary.

Stan, a former president of the Wyoming Stock Growers' Association, was the most intimidating person in a room full of personalities, who with the utmost urbanity and nonchalance expected to intimidate. Nonetheless, there was a hushed attention when he and Mary gave their presentation to the group, Stan struggling to temper his anger, Mary her grief. They had a personal authority, the cadence of truthfulness—that moment of danger and the daring leap toward it—that finds its mark and stirs even the most spin-addled among us. What else could I do but seek them out?

My Irish grandmother would have loved Stan, impressed above all by the badge of his Irish heritage, "a grand head of white hair." He looks like one of her seven brothers, all of whom were over six feet tall and wholly handsome. Coal miners in the Wyoming Valley of Pennsylvania, the brothers were physically powerful, silent survivors of the brutal labor conditions that marked the turning of the nineteenth century into the twentieth and agrarian America into an industrial behemoth. The Molly Maguires were one response to squalor and hopelessness, the labor unions another, President Theodore Roosevelt and his Progressive Era "square deal" politics yet another. And it was then, as the frontier closed and the industrial future opened, that most of the movements that we associate with a liberal society found themselves: conservation, labor rights, women's rights, child labor laws, racial equality.

It is a fluke of personal history that my mother was born late to my grandmother and left her own marriage early, so that I was mostly raised by my grandparents in "the hard coal," under the aegis of union solidarity and community empowerment. Growing up with them was about as wild as anything the West has to offer: watchful of the Susquehanna River at our front door, slow and gray until the flood seasons when it threw boulders down its length; mindful that the "bubble fountain" outside my cousin's house was really a sulfuric pit, the fumes from which could render unwary children unconscious; breathless with anticipation as my grandmother's well-worn hands lifted a robin's nest from the cherry tree to show us how it was lined with her soft white hair. These stories, shared around kitchen tables far from my grandparents' lives in time and place, allow me to participate in ranchers' lives as one of their own. When I tried to edit them from this story, Carol Hamilton protested: "Ranchers will read this and will want to know who you are. Besides, I liked knowing that someone from the East has a story not unlike my own." Without exception, ranchers point to my working-class, rural background as unusual among environmentalists, the missing, essential touchstone.

In *Forcing the Spring*, Robert Gottlieb sums up the complaints against the environmental movement's white, middle-class elitism as "the failure to incorporate equity or social justice considerations in selecting the issues they fight, and the disregard for local cultures and grassroots concerns." He opens his history of the environmental movement with an account of the 1991 People of Color Environmental Leadership Summit, where Dana Alston said that people of color did not want a "paternalistic" relationship with environmental organizations "but one based on equity, mutual respect, mutual interest, and justice." Stan Flitner recently discovered common ground with the East Coast's African-American community through his service as cochair, together with an African-American community leader from New Haven, Connecticut, on the Affected Communities Subcommittee at the U.S. Institute for Environmental Conflict Resolution, run by the Morris K. Udall Foundation in Arizona. The committee is tasked with finding ways to better implement the "early and effective involvement" of resource-dependent communities in the decisions that affect their survival. As I listened to Stan describe their work, it seemed to me that their first challenge was to make visible what has become invisible to environmentalists in the National Environmental Policy Act: that the human presence on the landscape cannot and should not be erased from it, that resource-dependent communities have rightful economic and cultural claims on place. A reminder, in short, that as Theodore Roosevelt IV said in a speech to the Boone and Crockett Club (the oldest hunting and conservation organization in the United States), "No matter what else we are talking about, environmentalists need to remember that most of the time we are also talking about *someone's home.*"

After his last trip east, Stan excitedly told me, "I learned a new term in New Haven: *gentrification*. It's exactly what we are facing," referring to the unprecedented turnover of working family ranches to wealthy hobbyists. The suspicion that rankles the rural West is that many environmental groups prefer the turnover: better to lose one large ranch owned by a surly westerner and gain thirty subdivisions with million-dollar second homes for "liberal outsiders" establishing residence in a state without personal income tax. How else, they say, can you explain examples such as the one in Jackson, where environmentalists are fingered as refusing to reach a compromise over the dire straits in which a well-loved ranch family found itself after a tragic personal loss, one that had a negative impact on their grazing leases in Teton National Park? Ultimately, the ranch went with the leases. There are only a handful of working ranches left now in Teton County, one of the wealthiest counties in the United States and the only county in Wyoming that went for John Kerry in 2004.

Just prior to the election, several environmental organizations pondered whether or not they should support the voluntary buyout of federal grazing leases, a proposal vigorously opposed by most in the ranching community. Given ranchers' vast oral history of betrayal at the hands of environmentalists, they naturally figure the buyout program would end up about as voluntary as federal income tax. They call it "ecological relocation" and compare it to the Chinese government's attempted removal of Mongolian herders from their ancestral homeland. Instead of helping the community survive and do better by the land, ranchers view this program as just more evidence of "enviros" seeking their demise, either slowly (as one wilderness advocate hopefully noted, "time is on our side"—i.e., ranching is headed for the economic graveyard anyway) or swiftly by a regulatory ban or court-ordered bullet. In fact, many rangeland scientists today point to the benefits and necessity of well-managed grazing on an ecosystem that evolved in the presence of large herbivores, not only bison but other large Pleistocene-epoch grazers. Tony Burgess, a rangeland scientist at the University of Arizona, says that the time has been so short since those Pleistocene herbivores became extinct that "the grasses don't know they are gone."

In response to her experience in TR's birthplace, Mary Flitner sent the organizers and me a letter: "Ranchers' long-term commitment to place has brought historical connection and longevity of knowledge of land, beyond boundaries and regardless of ownership. The continuity of this view is valid, deep, and not duplicated elsewhere." Pat O'Toole, a sheep and cattle rancher out of Savery, Wyoming, who served on President Clinton's Western Water Policy Commission, told me on this trip, "Our country is fast approaching a precipice; we are about to make decisions about water, trade, and land that will be ruinous to both rural communities and the landscapes we tend; decisions that I am afraid our nation will long regret." Perhaps the rest of us in the Blue States will be as blasé as the grass and not notice that these communities are gone. After all, we hardly notice they are here until, that is, a presidential election doesn't go our way.

All of the anywheres in Wyoming seem to be located at the far-flung ends of high desert basins, endless expanses of gray sagebrush or just plain dirt, with a few sometimes starkly obscene rock outcroppings (well, after hours of driving at eighty miles per hour, the mind wanders) that could have dropped from Mars or just been wished into being by the sheer mercilessness of the place, what John Wesley Powell, the explorer, so quietly described as the West's single unifying characteristic: aridity. As one Wyoming governor purportedly described it, "Wyoming is a small town with very long streets." And endurance drivers. On the verge of nodding off behind the wheel on Interstate 80, traveling from one ranch oasis (green fertile valleys) to another, I couldn't imagine how anyone had the temerity to call this landscape home. In what I would describe as a brazen homage to the frontier, Interstate 80 was built more or less along the course of the Overland Trail.

Carol Hamilton's family, the Byrnes, must have staked out the Overland Trail. As Rick Knight breathlessly whispered to me, "They came here before the Civil War. My god, that's early." They settled a town eventually named Piedmont, as Mr. Byrnes's wife was from Piedmont, Italy. No more than ten miles from Carol and Richard's ranch, down a switchback road in the middle of nowhere, today's Piedmont is a ghost town. A handful of gray timber structures—once homes, a bank, a telegraph office, and I hope for their sakes, a saloon—slowly falling back into the surrounding gray landscape. While Piedmont seems in most respects a classic ghost town, it wouldn't be Wyoming ranch country if there weren't a twist. At the town's edge, three large, domed stone structures stand in an intact, neat row. Much like the beehive monasteries of Ireland, these too are situated in a landscape that cries out for a good hermit or two. But in keeping with the ferocious practicality of the West, these are the Piedmont Charcoal Kilns, part of a once-thriving industry, processing charcoal for mining smelters. About these durable, surprising kilns, Carol said, "You know it was those Italian women who knew a thing or two about masonry; you know it was actually them who were responsible, though they didn't get the credit." Wrangled the town's name from its original designation—Byrne—to Piedmont, though, didn't they?

Wyoming was the first government in the world to give women the right to vote; it wasn't even a state then, just a territory. Maybe I heard that in one of my feminist history courses in college. Perhaps I forgot because it seemed so unlikely at the time. But watching Carol, who broke her hip last year (the year she turned sixty) in a riding accident, watching her ride—careening around trees and through gullies, storming over flats, running down and cussing with the manliest of them, and all despite a lingering intractable fear of horses, worse than the diminished flexibility and strength after the fall—well, you figure that Wyoming had no choice but to surrender up some suffrage. Cynthia Lummis, the state treasurer and a much-rumored future governor, hailing from one of Wyoming's oldest ranch families, once blithely told me, "People in Wyoming are just live-and-let-live sorts of folks." I burst out laughing. Cynthia, who has a steady, straightforward, normally undeterred approach to the world, paused. I explained: I laughed not because it's not true, but because it actually is.

Cynthia's sister runs the family ranch and recently barely survived what ranchers call a "wreck"—an accident generally involving a human, a horse, and/or a cow. The accident was so bad that the fate of the family ranch was in question. Cynthia said, "It's not a matter of *if* in this work, it's a matter of *when* and *how bad*." The stories come with the territory and the endless cups of coffee. There's the one about the rancher, a great horsewoman, who was working a bull in a corral, and the bull charged under her and the horse, lifted them both up, and threw them over; she and the horse were killed. Or there's the one about the rancher whose horse reared up after the rancher roped a cow and had the line dallied around his saddlehorn; when he let the rope go, it turned out the weight of the cow was the only thing holding that horse down; the horse fell over onto its back with the rancher under him; both horse and man were seriously injured. Or there's even the Hamiltons' notorious dog, probably the only dog in all of Wyoming with a Japanese name, who was stepped on by a cow, shattering a bone in one of her legs; with the pin protruding from the still-open wound, the dog still insisted on joining the roundup, making her jubilant way on three legs through territory I could barely negotiate on top of a horse. Dogs, horses, humans, all in life-and-limb jeopardy in this work, and yet all apparently in love with the work.

Baxter Black, the cowboy poet, explains that he writes for those who have discovered the myriad ways to be "bucked off, stomped, bit, run over, butted, hung up, pounded, pawed, drug, flattened, smoothed out, licked, whipped, stepped on, spit on, pooped on, peed on, calved on, shined on, fooled, tricked, out-smarted, buffaloed, horned, humped, or humiliated" by "semi-domesticated large herbivores." Baxter is a friend of Cynthia's, and she seated me next to him at the closing dinner of the Western and Midwestern State Treasurers' Conference in Jackson, Wyoming, which she was hosting. Perilously thin and fiendishly ironic, Baxter is also gallant, modest, and close to a modern-day Mark Twain. Sitting next to someone so acute, I felt myself ebbing into dimmer and duller. But what was far worse, I sat there like an eastern rube, entirely ashamed that I was unfamiliar with his work and only remotely knowledgeable about the lives that inspired it.

Most attribute the expression "radical center" to Bill McDonald, a rancher, cofounder of the Malpai Borderlands Group, and MacArthur

Genius Grant recipient. The community forestry movement calls it the "radical middle" and attributes it elsewhere. Bill would say that an idea whose time has come will be found by many of the thoughtful among us. Whatever its provenance, it's an idea that is gaining momentum in rural America. It has arisen from the conviction that pressure politics on both sides of the divide has reached its nadir, that those engaged in this conflict (environmentalists, property rights advocates, local communities) have become more consumed with winning the argument than with finding solutions, and that the landscape and its people cannot survive much more of this combat. The last political resort of those for whom anger no longer suffices (though it does still prick and prod), the radical center seeks to address the long-term survival of both agriculturalists and landscapes by focusing on solutions that encompass environmental stewardship and economic vitality as mutually reinforcing common and community goods. As Dan Kemmis writes, "The future of the West must involve a radical and permanent transcendence of the region's embedded struggle between imperial-type environmentalism and Sagebrush Rebellion–type resistance."

My problem: in the hands of the institutional class, this nascent, still tremulous, mostly grassroots movement may tend to ossify into ideology, into those moralizing postures of liberals, environmentalists, and conservatives that go against the grain of largely secular and pragmatic rural communities. Once ideas become owned by institutions—abstracted, formalized, separated—they risk becoming the tools of willfulness and power. In the wrong hands, lovely centrist words such as "collaboration," "trust-building," and the less than lovely zinger "transparent communication" become just a better way to control and manipulate, a sign that one group is special and owns the only right relationship to place.

Of these centrist signposts, "transparent communication" is the expression that consistently irritates me. It is a term lifted from the financial world, where its very impersonality stands as a guarantor of market discipline and reliability: that market information can achieve "perfection" and investors can achieve profits. In the world of normal human interactions, it claims far more than it can deliver—the notion that any of us, all so trained in social opacity, has even a moment of transparency (motives, failings, inadequacies all revealed). Simultaneously, it relieves those of us plagued with postmodern squeamishness over anything that implies judgment of the messiness and the obligation of the more personal expression "truth-telling": the antique idea that one fragile human trajectory could ever encompass anything so grand as the truth.

And then there are ranchers: with a practice of silence and habit of physical courage, they don't seem to have lost the muscle memory for truth-telling, however unvarnished or impolitic. As Carol Hamilton once said to me about environmentalists: "Talking with them is never about an exchange of ideas, and it's certainly not about standing in my viewshed and looking out at the world with me. No. You watch them lying in wait for you, their agendas in their back pockets, waiting for you to give them that key opening, that misspoken word." But I fuss. It's impossible to protect words from the clever, or bright ideas from the clumsily self-serving. The radical center movement looks like our best hope, if it can maintain a vigorous alignment with the rural communities that sired it, as well as a willingness for tough self-scrutiny.

The environmental community, though, has been dragged kicking and screaming toward any critical examination of its own biases, proving itself almost wholly resistant to reform. In a profound attempt at addressing this, William Cronon, a historian and environmentalist, published a collection of essays in 1996, *Uncommon Ground: Rethinking the Human Place in Nature*. Many environmentalists burst into flames upon reading that the category of *nature* does not stand above and beyond human interpretation in the sacred space of *Nature*. In one of the more controversial essays, "'Are You an Environmentalist or Do You Work for a Living?' Work and Nature," Richard White wrote:

> Most environmentalists disdain and distrust those who most obviously work in nature. Environmentalists have come to associate work—particularly heavy bodily labor, blue-collar work—with environmental degradation. ... Environmentalists so often seem self-righteous, privileged, and arrogant because they so readily consent to identifying nature with play and making it by definition a place where leisured humans come only to

visit and not to work, stay, or live…. Without an ability to recognize the connection between work and nature, environmentalists will eventually reach a point where they seem trivial and extraneous and their issues politically expendable.

What the environmental community concluded after the 2004 election, however, is not that their discourse, attitudes, or cultural biases with regard to the West and blue-collar rural communities require vigorous examination, but that rural communities require "conversion." When they want something from ranchers, such as support of wilderness designation on public lands, they send out envoys, whom I have been told speak "rancherese." This usually amounts to opening any conversation with "howdy" and assuming a laconic, relaxed tone. I'm not sure that any ranchers are taken in, as most feel that they do indeed speak English (sometimes Spanish or French or all three). Rural people recognize this for what it is; as one commented, "We want real partnerships, not partnerships of convenience."

There are some environmental groups that are now acknowledging that rural people don't much like them, but feel the dislike is just a perception problem on the part of rural people. Agriculturalists don't understand that the environmental *mission* is essential to their lives." It doesn't seem to occur to environmentalists to change the premise. Instead of saying to rural people, "What hurts the environment hurts you," perhaps the environmental community could realign itself with its socially progressive roots and find the capacity to say, "What hurts rural working people hurts the environment, and hurts us all." One environmental group defended its "community outreach" this way: "building new relationships using a combination of strong science and values-based messaging to talk about land protection in a way that is compelling and meaningful to a broader array of people who care about land conservation." The first question that comes to mind is *whose* values? Then whose science and why is *yours* "strong"? Then, why is the land's protection separated from that of the community? And finally, there is the fervent wish that environmental groups could wean themselves from marketing consultants and the glib, self-aggrandizing, and valueless spin that results. *These* are the "core rural values" that fly over the heads of most environmentalists: reciprocity and faithfulness to the community, what is required for a barn raising or a roundup.

Every evening, sometimes after a hot shower (stink usually needing to weigh heavily against exhaustion), Rick, Carol, Richard, and I sat down to beef and beer, or beef and wine, or beef and vodka, and uproarious conversations. Our suppers might have been mostly silent if it weren't for Rick. He is notorious in the ranching community for his staggering ability to talk and inability to do so on the subject of ranching without breaking down into sobs and tears. Jon Christensen put it this way in *High Country News*: "Rick's study on rangeland biodiversity has come to define the issue as much because of the personality and advocacy of the scientist as for his results."

Once, over a few beers in Denver where we had gathered yet again to try to improve the economic chances for ranchers, Bill McDonald (that rancher who won the MacArthur Genius Award) asked Rick, "Son, how is it that you can't get that crying thing under control?" Ed Marston, the former publisher of *High Country News*, once speculated that "Rick is mourning the loss of an America that could understand their [ranchers'] worth." That's a good answer, but Rick had another: mention ranching and his mind floods with the faces of all the ranchers he knows and loves, people whom we will not see the likes of again once this way of life vanishes.

On this trip to the Hamiltons', Rick fell upon an affectionate distraction for his own tears. Whenever he started to cry, he turned to me (knowing, without needing to check, that I was also crying) and teased: "Are you indulging yourself in that sentimental nonsense again? What are we going to do with you? Can't you get a grip?" The joking gave Richard Hamilton a break from his embarrassment. And that was the whole purpose. Gradually, Richard has grown more sanguine about the ways in which I embarrass myself. On my first trip to the ranch, when he, Casey, and I rode out to the mountains to put down salt for the cattle, he turned back to me gruffly at one point and said, "*Who* are you talking to?" I thought, "Not you, you cranky old bastard," but said what was even worse: "I'm talking to Chester and the dogs, *of course*" (ranchers don't routinely chat up their dogs). On this trip, Richard didn't even flinch when he heard me urge the cattle on with "Come on sweetheart, we are going in this direction now."

Over the years, Richard seems to have learned a form of patience that looks like forbearance but might actually be closer to the entertainment of irony. He has endless stories of range scientists, state wildlife officials, national wildlife officials, and writers like me stumbling over our own theoretical orientations to the land. Once, he encountered a wildlife biologist on his private property, where she and her students were conducting riparian studies under the mistaken notion that they were on public land. She pointed to the bare cutbanks and with great umbrage asked Richard if he "understood" that his cattle harmed the cutthroat trout that spawned there in the spring. I can see him now: slowly getting off his horse, his white hat, chaps, vest, and his measured, deliberate approach. With an authority born of years of hard, thoughtful living on one piece of land, he instructed the scientist on looking more closely at the signs and on understanding the particular history of an area. The banks of that stream had been "cut" by the old tie drives that occurred in the high-water season to bring railroad ties from the mountains downstream to the once-bustling town of Piedmont. He told her that the cattle aren't even near that stream in the spring, that he doesn't rotate them into that area until after the trout have spawned. He then showed her and her students around the ranch, including their ongoing work on public and private lands to restore riparian areas. She admitted to him later that she was taught in college that cattle and ranchers were an ecological evil, and she began with that assumption. She still writes to the Hamiltons.

On the last evening at the ranch for Rick, the four of us indulged in a bit more vodka and conversation than usual and talked about the roundup. I said how impressed I was with Casey, the main hand, at home that evening with his own family. Casey is perhaps thirty, rail-thin, quiet, and earnest. He is that great blessing of a human being: someone who is quite simply and entirely all that he should be. And that *all* is a prodigious quantity. Watching Casey ride, without spurs or bit (he uses a bosal), without fear or self-consciousness, and through the pain of several serious old rodeo injuries, well, watching him ride comes close to an act of prayer because his horsemanship seems so much an act of grace.

During the corral work, Casey was the master of the chute and I was his partner. The metal chute has three parts, operated by three levers, that require perfect timing to immobilize an animal safely, such that the cow doesn't hurt herself or the humans tending to her. I imagined that I could help Casey by operating one of the levers. He looked alarmed, but wanted to be polite. He didn't say, "Dear god, no. Are you out of your mind?" Instead he explained that, when one of the handles flies back, it could easily "knock you unconscious." I figured that was all Richard would need—an unconscious greenhorn—so I said, "Well, maybe not then." Casey didn't tell me that operating the chute is a one-person job, as that would have pointed out my deep ignorance of corral work. He would never deliberately embarrass me.

Once the cow was secured in the chute and the pregnancy tester determined that she wasn't barren, Richard inoculated her, Carol sprayed her with deworming fluid, and Casey removed the old ear tag with a hooklike device and waited for me to hand him the new one. Seconds before, he would call out to me the ear tag color—white, black, or yellow, each indicating a different age for the cow—to load into the stapling device. All of Casey's self was expressed in the way he called out those colors to me. I was his teammate, his partner. Never before have white, black, and yellow sounded as important as when Casey said them: as though he were handing off to me a vital task, elevating my small contribution to their efforts to equal standing with his own considerable skill. Along the way, he occasionally apologized unnecessarily for calling out the wrong color or for a momentary slowness, all the while never failing to nod his assurances to me when I made a mistake. Hour after hour we all worked this way: focused, tenacious, respectful of one another.

At the end of the day, as we ate our sandwiches and drank our beers leaning up against a pickup truck, I said to Casey: "My god, Casey, you are an inspiration. You work so hard and without any complaint." He replied: "It's not hard, Joan. You work the chute; you don't let the chute work you. Besides, I love what I'm doing." I thought of Wallace Stegner, whom Paul Starrs describes as the "patron saint of thinking environ-mentalism," and who once wrote about ranching, "I have known enough range cattle to recognize them as wild animals; and the people who herd them have, in the wilderness context, the dignity of rareness; they belong on the frontier, moreover, and have the look of rightness."

We ended our last bout of vodka-steeped, kitchen-table storytelling that night with Rick's recounting of his final conversation at the corral with Casey: how Casey thanked him for his help, and how Rick said that it was a privilege, that he took these experiences back to his students in wildlife biology and rangeland ecology. Rick asked Casey, "Is there anything you would like me to tell them?" Casey replied, "Just tell them that we are good people." The shame of it is that they require a messenger.

Drum Hadley is the author of four books of poetry, his last titled Voice of the Borderlands. *Here, he recounts the experience of hearing a developer considering ranchland as a commodity, as a "thing."*

Our Lands in the Belly of the Beast

Drum Hadley

Businessmen overheard in the Palm Court Restaurant, Plaza Hotel, New York City, America

Take them. Invite them out to lunch. Don't get attached.
It's money, it's property, it's real estate, it's things, you say.
Do you want to sell?
What do you want to pay me to sell it for you?
We made fifty or sixty million. He died penniless.
His children asked us for one thousand dollars for the funeral.
It was tax deductible. Of course, we gave it to them.
Is it legal? The government pays lip service,
But the competition is fierce. You will have no license.
You will be seen as a mortgage consultant, a foreigner,
Your English will not be like theirs.
The property will be worth more
Than eight to ten times what we pay for it.
You will ask them for three or four,
But before they know it, you will get ten million.
When I am going to foreclose,
I pay the lawyers, maybe fifty thousand,
Four percent, three percent, five percent.
The poor man never wants to pay the lawyers,
So you know the poor man will lose.

These companies are like candy stores.
Make all the dough you can. Buy today, sell 'em short.
Take them, invite them out to lunch.
Subdivisions in mountain valleys,
Wild running rivers, country ways of livelihood,
Deer, cowboys, mountain lions, javelinas,
Don't get attached. It's money. It's property.
It's real estate. It's things.
Take them; invite them out to lunch.

Although Julene Bair left the high plains of western Kansas when she was eighteen, she keeps going back for holidays and in her writing. Much that she once loved has disappeared. In this essay, she charts another eminent disappearance: the region's groundwater. Her book One Degree West: Reflections of a Plainsdaughter *won Mid-list Press's First Series Award and Women Writing the West's Willa Award.*

She Poured Out Her Own

Julene Bair

Used to be you could see the place from miles away—not only because my grandfather built a grand house in 1919, but because he chose the highest land around. "High Plains Farm," he painted in white letters on our red barn. Now all you can see is the silhouette of a pivot sprinkler.

About ten years ago the farmer who manages the land for the out-of-town owner bulldozed the house, the outbuildings, the yard trees, and the hundreds of elms in our windbreak. He burned the rubble piles and sold all the old implements to a scrap-iron dealer, leaving virtually no trace. He did this reluctantly. I know, because the day I discovered everything gone, he happened to drive up the dirt trail in his late-model white Ford pickup as I stood gazing in befuddlement at empty air.

Although this young farmer with the clean seed cap and thick red neck and arms, close-cropped hair and clean shave, was twenty years my junior and had himself grown up in a different county, he'd married into a Sherman County family, and through them knew the names of my grandparents and father and mother. I was the great-grandchild, he the great-great-grandchild of pioneers, and he could imagine what the old place must have meant to the Carlsons and Bairs. He was apologetic, though powerless. The abandoned farmstead was rendering useless a good flat quarter section that could be planted to corn, a Program crop.

Program crops receive government subsidies, practically guaranteeing a profit. I know this because my family still farms. In the 1960s, my parents traded this home place for land closer to their other holdings. Like many other successful farmers in our northwest corner of Kansas, they moved to town, where they built a brick ranch-style house that would blend well in any suburb. From then until his death in 1997, my father commuted to farm the land we still own.

Every few years, I obey the compulsion, as instinctual as a migratory bird's, to return to the home nest. Last time I went, I parked my car by the pole that used to bring electricity to our house but now conveys power to the pivot sprinkler, keeping it clocking around the field. Irrigation circles are a quarter-mile across, and the center of this one lay beyond where our north windbreak had been, not a great distance over the empty ground, but in my early childhood, anything beyond the windbreak had seemed like the edge of the earth. The sprinkler wasn't running that day, but sat stationary. If due north was noon, then the long line of its rigging and towers pointed to two o'clock, toward where the sheep barn had once stood. A thousand ewes and their lambs used to churn and bleat in the corral when penned there for lambing or to be dipped for ticks, vaccinated, and marked with chalk. They always ran in a circular motion, like soap bubbles at the bottom of a sink, but now it was as if all life, human and animal, had circled down the drain.

It was late May, and I walked down the rows of ankle-high corn examining the ground. I found a curved piece of white glass and a saucer shard with a faded orange flower painted on it. I examined both for a long time—this broken lip from the bowl my mother used to mix cakes in, this fragment of a plate my family and I used to eat from. I am always surprised by how a small scrap of the past can excite me, how alive the connection still is. Any discovery at all is like having a lucid dream, a direct link to the revelatory power of the subconscious. Along with the object, restored in memory, bloom images of my mother and father in their prime and all the life their union and work brought into being, including younger versions of my brothers and myself.

The object that most arrested my attention that day was the head of our old windmill. I found it lying on the ground in a tall clump of weeds near our pit silo, where we used to burn our trash and throw our junk. I must have seen it on previous visits, but in waking life, as in dreams, we fasten on those objects that have immediate meaning for us. I had recently begun reading, thinking, and writing about the Ogallala Aquifer, and I could not imagine a more telling artifact.

The windmill head lay in a nest of bent vanes, some of them buried almost completely in the dirt. Inspired to take a rubbing of the embossed print on the gearhead, as from a gravestone, I went to my car and grabbed a pencil and a piece of paper. Back at the mill, I had difficulty holding the paper flat as the wind whipped the edges. It would have been an unusual day on the plains had the wind not been blowing. Even on relatively quiet days the well-greased reel had turned, the shadows of the vanes serenely revolving over the ground in my mother's vegetable garden, irrigating our yard as well as providing water for us and our livestock. Now, in an odd-seeming role reversal, my shadow fell across the rusted crankcase and twisted vanes of the bodiless giant. I rubbed the edge of my pencil back and forth.

FAIRBURY WINDMILL CO.
FAIRBURY NEBR 10-34
PATD. DEC. 04, 1926

I used to climb the tower. I liked to sit up there and gaze over our windbreak at sunset. It delighted me to think how magnificent the prairie must have been in its original state, to imagine buffalo instead of our sheep grazing the pastel hills above the Little Beaver, the dry, sandy streambed that meandered northeast through our sheep pasture, headed toward its outlet in the Republican River beyond the Nebraska border. I was always careful to brake the rotor before climbing the mill. I wouldn't want to be sitting on the platform when the fan began turning, and even on quiet evenings, I never knew when a breath of wind might arise. The latent danger demanded my respect.

Now the tallest, most commanding object on our farmstead lay at my feet. The windmill's corpse reminded me of the buffalo bones that

I'd read had littered the plains when settlers first arrived, on the heels of the hide men and soldiers. All that waste. It reminded me also that the destruction had never ceased.

As the historian Walter Prescott Webb pointed out in his seminal 1930s history of the Great Plains, until immigrants from Europe, the eastern United States, and the Midwest were able to shed their preconceptions and meet our treeless and virtually waterless savannah on its own terms, it remained the Great American Desert the explorer Stephen H. Long had labeled it. The incessant wind blew many pioneer families back home before those who remained recognized its value, but gradually, homestead by homestead, as Webb put it, "primitive windmills, crudely made of broken machinery, scrap iron, and bits of wood" began to appear. These "were to the drought-stricken people like floating spars to the survivors of a wrecked ship."

I have read that the Apache and Pueblo Indians emerged into their desert by way of sacred springs. We came into ours up the stems of windmills. I'm speaking figuratively, of course. The literal movement was downward, into the wells. Some of the earliest settlers hand-dug their wells, a dangerous enterprise, as the sides could cave in. Later, horses were employed to turn augers and then to hold the weight of casings as they were lowered. One pioneer descendant, quoted in the Sherman County history volume *They Came to Stay*, remembers her father being eased down a well on a swing with a sledgehammer in his hands. His job was to steady the casing while the men on the surface clinched the rivets connecting the next section. "I visioned the rope breaking or the case slipping," recalled the daughter. "I was in control of the horses holding all that weight! As it got heavier, my horses began to strain." By the time a fresh team was brought, hers "were stretched on their bellies."

This firsthand account reminds me of the pioneers' heroic accomplishments and perseverance, but lurking behind the story is an unremarked miracle, the water itself.

In 1899, when the geologist N.H. Darton named the rock formation containing the Ogallala, he was probably thinking of the southern

Nebraska town of Ogallala and not the Oglala Sioux Indian tribe that once occupied the region along with other Plains tribes, most notably on my west-central plains, the Arapaho and Cheyenne. He may not have known the meaning of the word in Sioux, which I have seen variously translated as "to scatter one's own," "she poured out her own," and "spread throughout." Yet no name could have been more appropriate. The water in the Ogallala is itself spread throughout the area its tribal namesake once roamed, all the way from South Dakota to the Texas Panhandle, 174,000 square miles.

If current-day plains dwellers and interested ex-natives like me were mindful of our bioregion, we would call ourselves the Ogallala people, not directly after the tribe that lost its home to the Euro-American invasion of the plains, but like this:

> the tribe
>   the town
>   the rock
>   the water

The name would cascade as water does, down stairs of years, onto us.

I never heard the words "Ogallala" and "aquifer" in my childhood. Water was water. Even though it was hard to come by, no one in my family nor any of my teachers dwelled on the science or mystery of its origins. Not until enthusiasts began promoting irrigation did the word "aquifer" enter the plains farmer's lexicon, as a limitless underground lake. In the aftermath of the 1950s drought, the notion of engineered rain from a source so plentiful it would never run dry must have cheered farmers up quite a bit. The promise must have seemed like the fulfillment, finally, of the fantasy proliferated by earlier settlement boosters who promised that rain would follow the plows as pioneers moved westward.

If we had been aware of ourselves as belonging to a cherished place, one that we wished to leave intact for future generations, we might have reacted to the promises of irrigation promoters the way the Hopi Indians did in the 1970s, when the Bureau of Indian Affairs wanted to drill a well and install a water tower in Hotevilla, on Arizona's Third Mesa. Their ancestors, known to us as the Anasazi, had been desert dwellers for several thousand years. The knowledge that flowed into them from these deep roots was far more persuasive than the government hydrologist's promises. The elders reasoned that the ability to store thousands of gallons of water in the tower would engender a false sense of plenty. They knew that the tower would lead to waste and the pumping would dry out the spring where they'd gathered water for centuries. I've since read that the elders lost the battle against the tower. It was installed, but most people in Hotevilla refused to hook their houses up to it.

The pioneers and we descendants have always acted in the opposite manner, grasping whatever new technology would make our lives easier. As justification, we point to the hardships we faced. When I dared to ask one of my father's old sheep buddies, as they liked to call each other, if he regretted our having plowed most of the prairie under, he said, "Hell, no." This man, like my father, had grown up in a sod house. "We had it hard. Baloney—good ol' days. Outside toilets, freezin' your butt off. Look at you, Julie. You're sittin' in a pretty nice chair; you're not out in a teepee somewhere, weavin' wool."

Despite cold and a host of other hardships, the Indians, whether they wove wool in southwestern hogans and adobes or tanned buffalo hides in plains tepees, developed a different set of values. The Hopi are actually grateful they live in a desert. In too easy a climate, they were told by their maker, Spider Woman, they would fall into ignorance and irreverence, as they had in other incarnations in the previous three worlds they inhabited. The unreliability of the rains keeps them diligent in their rituals.

Without a spiritual tradition that recognizes the balance of nature and holds it sacred, our relationship to the land and its bounty is like a child's in a candy store with no adult present to restrain us from gorging. We don't identify ourselves as natives of ecosystems bounded by natural limits of land and climate but as citizens of countries, states, and counties, and as owners of farms, places demarcated by lines on maps. We conduct ourselves within an economy that depends on the depletion and degradation of the real things—plants, animals, soils, air, water—that sustain us.

Today, people know that the Ogallala Aquifer is not a lake but a vast accumulation of gravelly deposits saturated with fossil water. These sediments washed down off the Rockies at the end of the Miocene epoch, five million years ago, and were deposited by streams that changed course continually, braiding, unbraiding, and rebraiding themselves over the plains. Glacier melt from the mountains collected in the deposits, but today little water from the mountains reaches the flatlands. The aquifer depends mostly on rainfall for recharge, and in the dry climate, most rain evaporates or is used by plants before it has a chance to seep downward. Today's sprinkler irrigation systems are about twice as efficient as the gated pipe we flooded fields with when we first started irrigating, yet the most recent Kansas Geological Survey reports a yearly pumping average of seven and a half inches, fifteen times more than the most optimistic recharge estimate of one-half inch.

It has been only forty years since "development," or mining of water, became widespread in my region, but depletion rates are alarming. I have become accustomed to the funnel shape of the hydrology maps: wide at the top, where the aquifer underlies most of Nebraska, narrowing to a rounded tip just south of the Texas Panhandle. The shape is reminiscent of a whirlwind or dust devil, a common sight on high plains fields, or the actual, more or less conical shape of a heart.

On the Kansas Geological Survey map, the western half of Sherman County where my family still farms is mostly solid orange, indicating declines of 20 to 30 percent. In southern and central Sherman County, Rorschach blots of brighter orange show depletion rates of 30 to 45 percent. A couple of counties south of ours, large areas range from dark brown to almost black, for declines between 45 and greater than 60 percent. Studying maps from different years, I have seen how such dark areas of high decline in Oklahoma and Texas began as little freckles but spread like cancers into oblong or larva-shaped blots. These gradually enlarged, indicating whole regions depleted below usable levels.

Our windmill pumped water onto my mother's garden and into our house and stock tanks at the rate of thirty gallons per minute. Since my father's death, I have been preparing our family's water reports and have acquired a disquieting awareness of today's irrigation rates, which range from 500 to 1,200 gallons per minute, and of how much water that equals over the course of a growing season. Each summer our farm's five irrigation wells pump between 100 and 300 million gallons. Sherman County's 886 wells pump between 29 and 50 *billion* gallons. By comparison, the city of Denver sends 70 billion gallons through its pipes. In one dry year, Sherman County's 158 irrigation farmers used two-thirds as much water as used by the 1 million people served by the Denver system. And ours is just one of several dozen high plains counties where irrigation farming predominates.

Farmers must file annually with the Kansas Water Office, reporting how much water they've used, but they have political clout and have so far resisted any serious curtailment of their water rights. Some minor restrictions have been passed. Existing rights have been frozen, no new rights will be issued, and a recent regulation requires all irrigators to install meters, making it more difficult for them to underreport the number of gallons they pump. But to date, the serious conservation measures proposed by directors of water control boards and state governors have not floated.

A plan called Zero Depletion, for instance, had the worthy goal of "sustainable yield" and would have set a future date after which no more water than what seeped into the aquifer would have been pumped out of it. One of our current farm's neighbors, whose father worked for my grandfather as a young man, tells me, "If that Zero Depletion had gone through, you could have shot a bullet down Main Street and not hit anybody." He is probably right. Plains economies depend on irrigation. But unfortunately, he was describing an inevitability, whether the plundering is stopped by regulation or depletion. As the desert writer Charles Bowden puts it, "Humans build their societies around consumption of fossil water long buried in the earth, and these societies, being based on a temporary resource, face the problem of being temporary themselves."

My father's old sheep buddy thinks he sees the writing on the wall. "You know, Denver's gettin' so huge; where in the world are they going to get their water?"

"Where's California going to get *their* water?" his wife put in.

Her husband resettled his cowboy hat, which he'd placed on his knee when he sat down to talk to me in my mother's living room. "That's where the water trouble's going to come from, because all these legislators in the cities want water for their people. They're not going to worry much about us out here gettin' a little water or not. They're going to try to tie up all the water they can."

His wife said, "I don't think the people in the cities have any idea about how important farmland is and what the farmers are doing. They're not going to think about it and realize the importance of it until they look and the shelves are empty at the grocery store."

When the environmental soundness of a practice is questioned, farmers and the ag industry often make this familiar argument. "How else would we feed the world?" they ask. The implication in this case was that if city people vote to curtail large-scale irrigation to secure their own water needs, a food shortage will result and people will be hungry. But as George Pyle, an editorial writer from western Kansas, argued in his book, *Raising Less Corn, More Hell*, the United States and Europe actually overproduce and undersell grain. They flood world markets with cheap, subsidized commodities. The shortage to worry about is in cash. More than one billion of the world's people earn less than one dollar per day and can't afford to buy enough food to eat even at the lowest prices. Like representatives of many countries who argue the point in the World Trade Organization, Pyle believes that U.S. and European farm subsidies exacerbate poverty and hunger in less developed nations, where farmers are forced off the land because they cannot compete in the artificially suppressed market.

Even if by some stroke of altruistic genius we came up with a means of feeding the world's starving without charge, we could do so most efficiently if we grew corn for human use instead of raising feed corn to fatten cattle and hogs. Over half of each year's corn crop is fed to livestock within the United States, as is much of the corn we export. By processing seven pounds of corn through a steer, we produce only one pound of weight gain, mostly in fat.

The chief irony in this waste is that we needn't have disturbed one blade of grass in the first place. The most convincing study I've read estimates that the nation's grasslands once supported around forty million bison. The environmentally exhaustive practices that support the beef industry could have been almost completely avoided had we stewarded instead of decimated the original herds. We might not be eating as much red meat, but we would be eating meat better for us.

And were it not for the government Farm Program, we would not be growing much corn west of the hundredth meridian—the invisible rain curtain running through the middle of the plains states and separating the high plains from wetter, more easterly regions. The crop requires two feet of moisture, about a foot more than what falls out of the sky in Sherman County in the ideal growing season. On one 120-acre irrigated circle, where a sprinkler equipped with the latest conservation technology applies water even at an ideal efficiency of 90 percent, this translates into about forty-eight million gallons. The Ogallala waters one-third of the nation's corn crop, and irrigated corn receives roughly that proportion of the $4.5 billion in annual corn price supports.

"They can keep their cheap food policy," said my father's sheep buddy. "We sold our souls when we started taking subsidies. If the government got out of farming now, we'd all go broke." The "cheap food policy" he referred to is farmer code for the government program designed to secure an inexpensive food supply. Government-subsidized grain suppresses the price that citizens, when wearing their consumer rather than taxpayer hats, pay and that farmers receive for grain. To increase both their income from the sale of grains and their share of Farm Program payments, farmers turn to yield-enhancing chemicals, genetically modified seed, bigger farms, bigger, and more complex machinery—and intensive irrigation. They have no choice, given the low price of grain, but to maximize their returns through increases in production and scale.

I asked our neighbor, the one whose dad worked for my mom's dad, what he would do to change things if he were boss of it all. "I don't really want to tell others what to do," he replied with typical plains humility, then proceeded to outline his own four-year crop rotation system: one season of pinto beans, two of wheat, one of corn planted back into the wheat stubble to conserve moisture. The system relied less on water-thirsty corn, but more on "no-till," a method that keeps weeds down with chemicals while reducing the number of passes a tractor makes through a field. He told me that Roundup, the chemical used more often in no-till than in conventional tillage, "is only contact. It doesn't go into the ground." I hope he's right, because as farmers are becoming more aware of the water's limits and as the cost of fuel to pump the wells escalates, they are turning increasingly to this alternative.

But environmental thinkers looking at the bigger picture suggest that, in the face of our dwindling fuel and water supplies, with farm chemicals showing up in our drinking water and with nitrate runoff killing the coral reefs and threatening life in the oceans, we can no longer afford to underwrite agriculture as currently practiced. The emphasis needs to shift to conservation: less rather than more reliance on chemicals; the restoration of our grasslands; more direct marketing systems that do not waste resources in shipping, processing, and packaging; smaller, biologically diverse farms; and the return to dryland agriculture on the plains. Given increasing global competition, farmers may always require subsidies to stay in business, but the only practices that warrant taxpayer support are those that truly do secure our food supply and those that preserve the land, water, and soil, not those that waste or pollute resources essential to the nation's future.

Instead of lobbying for revisions in the "cheap food policy," farmers are too often duped into blaming environmentalists for their problems. When I worry about chemical residue building up in our soil and water, or mention my regrets over the depletion of the aquifer, or suggest that the Farm Program should underwrite conservation rather than depletion, our farm's neighbors like to josh me about being a "greenie."

"We're the endangered species," says the neighbor with the no-till plan. He and another farm neighbor spend winter nights plotting their vengeance. "We're thinking of starting an adopt-a-prairie dog program for city folks. We could send them a picture each month. 'What Your Prairie Dog Did Today.'" He cracked a lopsided smile. "They could all be the same picture."

I had to laugh. But I also had to ask if he'd seen many burrowing owls lately. The odd little birds used to stand like sentries on our pasture's prairie dog mounds, but I'd read they were now rare.

He continued to grin. "We try to take care of them too, because they and prairie dogs go together. They're hard to shoot, though." He made a wavelike motion with his hands. "'Cause of how they fly."

He was both kidding and not kidding, I knew—getting my goat with the truth.

For my people the highest value has always been production and yields, the unbridled use of whatever could advance these, the removal or suppression of whatever got in their way. Yet my grief over the loss and destruction is not, as my plains friends assume, born of my life among urbanites. The environmentalists I met in college or in Colorado, where I now live, did not brainwash me. My conviction in these matters comes from my past on that farm. Had I not sat atop our windmill and gazed over what still remained then of the native buffalo grass, I would have no direct sense of what has been lost. I don't know a farm kid who didn't climb his or her family's windmill and ponder the same things. What we saw sank into us. Most plains-born people are not content in other landscapes. When I lived in the Midwest, going back to school, I hated not being able to see far, the humidity, the low overcast skies, and the blatantly green grass. The term "greenie" attaches the wrong hue to my environmentalism. I like the shortgrass prairie as much in its winter-cured, yellow phase as in the summer, when the pale, variegated greens range into blues. The plains are too intensely green now, almost every inch of the native grass gone, the sod turned and planted to nonsustainable crops made possible by wasting a substance to which we owe our own lives.

The water allowed us to live safely within shelterbelts and comfortably on lawns, the fragrance of domestic blooms floating around us, but we were also touched by wildness. June bugs slapped against our screen doors on summer nights. Toads hopped across our porches to feast on them. Lizards skittered through yucca litter in our pastures. The ears of kit foxes sailed over ditch weeds. Coyotes yipped from beyond the corral fences. Jackrabbits zigzagged drunkenly ahead of our cars. Prairie dogs and burrowing owls perched on our pasture hills. We would not have known these creatures had the water not made it possible for us to live where we did and to, by consequence, become who we were, with our particular sense of aesthetics, definitions of beauty particular to that place. We wouldn't have known the luminescent, high evening skies, the glorious sunsets over wheat fields and pastures, the soft pastels of buffalo and grama grass, the brilliance of snow-covered fields. We would not be us.

FAIRBURY WINDMILL CO. From my father's point of view, windmills were mechanical contraptions. He complained of the damnable amount of attention and servicing they required. The pump leathers had to be pulled and replaced every so often, the towers climbed, and the gears greased. Yet to most of us, windmills are romantic western icons. They stand starkly on the remaining plains grasslands. They seem to grow out of the ground like huge daisies, as if they were natural features or emblems of humans as natural creatures.

One family, one tower, and some danger in getting what that family needed from the earth in order to survive. The difficulty constituted what I have heard called "right relationship." Labor is expended, risks incurred that keep the supplicants mindful of their dependence on a resource, and the resource is not depleted, at least not seriously or rapidly. It is only tapped.

The rubbing still hangs over my desk. On its backside is a list of my mother's certificates of deposit, the only piece of paper I could find in my car that day. The irony doesn't escape me. It didn't escape me then. Since my father's death I had been helping her shop for competitive interest rates and keep track of her savings, money that had accumulated thanks to the Ogallala. Up the metal stems of windmills had flowed the water that made it possible for my family to establish a foothold, then a stronghold, from which we further enlisted that resource for our personal benefit. For thirty-eight seasons now, the water has gushed out of our wellheads. First, when we were flood-irrigating, we channeled it down the furrows of row crops ranging from sugar beets to corn and pinto beans. Now we sprinkle it on from overhead, as if it were real rain. When harvesting these irrigated crops, we have been harvesting the water, transferring it from the aquifer into our own dark bank vaults. In inverse relationship to the drawdown of the underground water, the money grows in storage, although there is really no vault and no sheaves of bills.

Our words for money come from actual things. "Buck" was originally the name given to a deerskin, a common unit of trade during this country's settlement. "Fee" comes from the German *vieh*, meaning "cattle." We use the term "shell out" because Native Americans traded in shells. "Salary" comes from the Latin word for salt, because Roman soldiers were paid partially in this essential mineral. But today our financial system rests on several levels of abstraction. The more years that separate us from the days when all of us ate directly from land and soil—when we ate our own grains, dairy products, vegetables, produce, and meat instead of the processed, pulverized, packaged foodstuffs they are now turned into elsewhere—the higher we have built the tower. We've removed the supports as we built, so that today our system floats on invisible perceptions. No stockpiles of gold back the dollar anymore. As the economist Milton Friedman explains, "pieces of green paper have value because everybody thinks they have value." Most transactions don't even require greenbacks. Our wealth is in name only, figures recorded, except on those occasions when we print them out, in binary code on computers.

My mind reels at how this transfer took place on the plains. We went from actual wealth in the form of natural resources on which all past and future generations depend to the individual abstract wealth of a few generations of pioneers and their descendants. Actual substance you can touch—real water from within real ground—has been transformed into binary code. We can't transform any of it back.

Wallace McRae is a Montana rancher and a longtime featured poet at the annual National Cowboy Poetry Gathering in Elko, Nevada. In this poem, he wonders about a world moving too fast to value what's close at hand and all that's gone before.

Things of Intrinsic Worth

Wallace McRae

Remember that sandrock on Emmels Crick
Where Dad carved his name in 'thirteen?
It's been blasted down into rubble
And interred by their dragline machine.
Where Fadhls lived, at the old Milar Place,
Where us kids stole melons at night?
They 'dozed it up in a funeral pyre
Then torched it. It's gone alright.
The "C" on the hill, and the water tanks
Are now classified "reclaimed land."
They're thinking of building a golf course
Out there, so I understand.
The old Egan Homestead's an ash pond
That they say is eighty feet deep.
The branding corral at the Douglas Camp
Is underneath a spoil heap.
And across the crick is a tipple, now,
Where they load coal onto a train.
The Mae West Rock on Hay Coulee?
Just black-and-white snapshots remain.
There's a railroad loop and a coal storage shed
Where the bison kill site used to be.

The Guy Place is gone; Ambrose's too.
Beulah Farley's a ranch refugee.
But things are booming. We've got this new school
That's envied across the whole state.
When folks up and ask, "How things goin' down there?"
I grin like a fool and say, "Great!"
Great God, how we're doin'! We're rollin' in dough,
As they tear and they ravage The Earth.
And nobody knows … or nobody cares …
About things of intrinsic worth.

In the late 1990s, former Sierra Club activist Courtney White dropped out of the "conflict industry" and decided to build bridges among ranchers, environmentalists, and others. In "The New Ranch" he takes us on a tour of a key moment in his journey.

The New Ranch

Courtney White

Ranching is one of the few western occupations that have been renewable and have produced a continuing way of life.
                                                                —Wallace Stegner

It had been a bad year to be a blade of grass. In 2002, the winter snows were late and meager, part of an emerging period of drought, experts said. Then May and June exploded into flame. Catastrophic crown fires scorched over a million acres of evergreens in the Four Corner states, making it a bad year to be a tree too.

The monsoon rains failed to arrive in July and by mid-August hope for a "green-up" had vanished. The land looked tired, shriveled, and beat-up. It was hard to tell which plants were alive, dormant, or stunned, and which were dead. One range professional speculated that perhaps as much as 60 percent of the native bunchgrasses in New Mexico would die. He looked gloomy. It was bad news for the ranchers he knew and cared about, insult added to injury in an industry already beset by any number of seemingly intractable challenges.

For some, it was the final blow. Ranching in the American West, much like the grass on which it depends, is struggling for survival. Persistently poor economics, tenacious opponents, shifting values on public lands, changing demographics, decreased political influence, and the temptation of rapidly rising land values for development have all combined to push ranching right to the edge.

If the experts are correct—that the current drought could rival the decade-long "megadrought" of the 1950s for ecological, and thus economic, devastation—then the tenuous grip of ranchers on the future will be loosened further, perhaps permanently. The ubiquitous "last cowboys," mythologized in a seemingly endless stream of tabletop photography books, could ride into their final sunset once and for all.

Or would they?

After all, for at least the past sixty-six million years grass has always managed to return and flourish. James Ingalls, U.S. senator from Kansas (1873–1891), and father of *Little House on the Prairie* author Laura Ingalls Wilder, once wrote: "Grass is the forgiveness of nature—her constant benediction. Fields trampled with battle, saturated with blood, torn with the ruts of cannon, grow green again with grass, and carnage is forgotten. Streets abandoned by traffic become grass grown like rural lanes, and are obliterated; forests decay, harvests perish, flowers vanish, but grass is immortal."

Few understand these words better than ranchers, who depend on the forgiveness of nature for a livelihood while simultaneously nurturing its benediction. And like grass, ranching's adaptive response to adversity over the years has been patience: to outlast its troubles. The key to survival for both has been endurance, the ability to hold things together until the next rainstorm. Evolution favors grit.

Or at least it used to.

Today, grit may still rule for grass, but for ranchers it has become more hindrance than help. "Ranching selects for stubbornness," a friend of mine likes to say. While admiring ranching and ranchers, however, he does not intend his quip to be taken as a tribute. What he means is this: stubbornness is not adaptive when it means fighting new knowledge, technology, and values in a rapidly changing world.

This is where ranching and grass part ways ultimately—unlike grass, ranching is not immortal.

Fortunately, a growing number of ranchers understand this and are embracing new ideas and methods, with the happy result of increased profits, restored land health, and repaired relationships with others.

I call their work "the New Ranch," a term coined years back in a presumptuous attempt to describe a progressive ranching movement emerging in the region. A New Ranch for a New West, that sort of thing.

But what did I mean exactly? During that summer of fire and heat I took a drive to see the New Ranch up close. I wanted to know if ranching would survive this latest turn of the evolutionary wheel. Was it still renewable, as Stegner observed, or is the Next West, a term I prefer, destined to redefine a ranch as a mobile home park and a subdivision? I wanted to see the shape of the future and, with a little luck, find my real objective—hope—which, like grass, is sometimes required to lie quietly, waiting for rain.

The James Ranch
North of Durango, Colorado

One of the first things you notice about the James Ranch is how busy the water is. Everywhere you turn, it seems, there is water flowing, filling, spilling, irrigating, laughing, moving. Whether it is the big, fast-flowing community acequia, the noisy network of irrigation ditches, the deliberate spill of water on pasture, the fish ponds, or the low roar of the muscular Animas River, take a walk in any direction on the ranch during the summer and you are destined to intercept water at work. It is purposeful water too, growing trees, refreshing chickens, quenching cattle, raising vegetables, and above all, sustaining grass.

All this energy is no coincidence: busy water is a good metaphor for the James family.

The purposefulness starts at the top. David grew up in southern California, where his father lived the American dream as a successful engineer and inventor, dabbling a bit in ranching and agriculture on the side. David attended the University of Redlands, where he majored in business, but cattle got into his blood and he spent every summer on a ranch. David met Kay, a city girl, at Redlands and after getting hitched, they quickly agreed to a plan: raise a large family in a rural setting.

In 1961, they moved to a ranch north of Durango and got busy raising five children and hundreds of cows. Before long, David secured a permit on nearby public land and began to manage his animals in the manner in which he had been taught: uncontrolled, year-round, continuous grazing.

"In the beginning, I ranched like everyone else," said David, refer-ring to his management style, "which means I lost money."

David followed what is sometimes called the "Columbus school" of ranching: turn the cows out in May, and go discover them in October. It often leads to overgrazing, especially along creeks and rivers, where animals like to linger. Plants, once bitten, need time to recover and grow before being bitten again. If they are bitten too frequently, es-pecially in dry times, they can die. Since ranchers often work on a razor-thin profit margin, it doesn't take too many months of drought and overgrazing before the bottom line begins to wither too.

Grass may be patient, but many bankers are not.

Through the 1970s, David's ranchlands, and his business, were on a downward spiral. "I thought the answer was to work harder," he re-called, "but that was exactly the wrong thing to do." Slowly, David be-came aware that he was depleting the land, and himself, to the point of no return. By 1978 things became so desperate the family was forced to sell a portion of their property, visible from the highway today as a res-idential subdivision called "The Ranch." It was a painful moment in their lives. "I never wanted to do that again," said David, "so I began to look for another way."

In 1990, David enrolled in a seminar taught by Kirk Gadzia, a cer-tified instructor in holistic resource management—a method of cattle management that emphasizes tight control over the timing, intensity, and frequency of cattle impact on the land. "Timing" means not only the time of year but how much time, often measured in days (as op-posed to months), the cattle will spend in a particular paddock. "Intensity" means how many animals are in the herd for that period of time. "Frequency" means how much time the land is rested before the herd returns. All three elements are carefully mapped out on a chart, which is why it is often called "planned grazing."

It has other names: timed grazing, management-intensive grazing, rapid rotational grazing, short-duration grazing, pulse grazing, cell graz-ing, and the "Savory system"—after the Rhodesian biologist who came up with the basic idea. Observing the migratory behavior of wild graz-ers in Africa, Allan Savory noticed that nature, often in the form of predators, kept animals on the move, which gives plants time to recover.

He also noted that too much rest was as bad for the land as too much grazing. In dry climes, one of the chief ways old and dead grass gets recycled is through the stomachs of ungulates, such as deer, antelope, bison, and cattle. Fire is another way to recycle grass, though this can be risky business in a drought. Either way, Savory's insight was to manage animals in a manner that resembled nature's model of herbivory.

As David and Kay James will tell you, however, the most important lesson they learned from Kirk was that their problem wasn't with their cattle management. What was lacking was the proper *goal* for their business.

"We really didn't have a goal in the early days," noted David, "other than not going broke."

To remedy this, the entire James gang (David thinks he might be related to the famous outlaw) sat down in the early 1990s and composed a goal statement. It reads:

> *The integrity and distinction of the James Ranch is to be preserved for future generations by developing financially viable agricultural and related enterprises that sustain a profitable livelihood for the families directly involved while improving the land and encouraging the use of all resources, natural and human, to their highest and best potential.*

It worked. Today David profitably runs cattle on 220,000 acres of public land across two states. He is the largest permittee on the San Juan National Forest. Using the diversity of the country to his advantage, David grazes his cattle in the low (dry) country only during the dormant (winter) season; then he moves them to the forests before finishing them on the 400 acres of irrigated pasture of the home ranch.

That's enough to keep anybody incredibly busy, of course, but David complicates the job by managing the whole operation according to planned grazing principles. Maps and charts cover a wall in their house. But David doesn't see it as more work. "What's harder," he asked, "spending all day on horseback looking for cattle scattered all over the county, like we used to, or knowing exactly where the herd is every day and moving them simply by opening a gate?"

It's all about attitude, David observed. "It isn't just about cattle, it's about the land. I feel like I've finally become the good steward that I kept telling everybody I was."

And the goals keep coming. Recently, the family developed a vision for their land and community 100 years into the future. It looks like this:

Lands that are covered with biologically diverse vegetation
Lands that boast functioning water, mineral and solar cycles
Abundant and diverse wildlife
A community benefiting from locally grown, healthy food
A community aware of the importance of agriculture to the
    environment
Open space for family and community

And they have written out the lessons they have learned over the past dozen years:

Imitating nature is healthy
People like to know the source of their food
Ranching with nature is socially responsible
Ranching with nature gives the rancher sustainability

Like the busy water at work on their home ground, David and Kay know exactly where they are headed.

It didn't stop there. Years ago David and Kay told their kids that in order to return home, each had to bring a business with them. Today, son Danny owns and manages a successful dairy operation that he began from scratch on the ranch; son Justin owns a profitable local barbeque restaurant; daughter Julie and her husband John own a successful tree farm on the property; and daughter Jennifer and her husband grow and sell organic vegetables and plan to open a guest lodge nearby. Only one child, Cynthia, has flown the coop, though fittingly she runs the charitable giving program at a major corporation.

In an era when farm and ranch kids are leaving home, not to return, what the James clan has accomplished is significant. Not only are the kids staying close, they are diversifying the ranch into sustainable businesses. Their attention is focused on the Next West, represented by Durango's booming affluence and dependence on tourism. Whether it is artisan cheese, organic produce, decorative trees for landscaping, or a lodge for paying guests, the next generation of Jameses has their eyes firmly on new opportunities.

This raised a question. The Jameses enjoy what David calls many "unfair advantages" on the ranch—abundant grass, plentiful water, a busy highway, and close proximity to Durango—all of which contribute to their success. By contrast, many ranch families do not enjoy such advantages, which made me wonder: what lesson can the James gang teach us?

I posed the question to David and Kay one evening.

"The key is community," said Kay. "Sure, we've been blessed by a strong family and a special place, but our focus has always been on the larger community. We're constantly asking ourselves, 'What can we do to help?'"

Answering their own question, David and Kay James decided ten years ago to get into the business of producing and selling grass-fed beef from their ranch.

Grass-fed, or "grass-finished" as they call it, is meat from animals that have eaten nothing but grass from birth to death. This is a radical idea because nearly all cattle in America end their days being fattened on corn (and assorted agricultural by-products) in a feedlot before being slaughtered. Corn enables cattle to put on weight more quickly, thus increasing profits while also adding more "marbling" to the meat—creating a taste that Americans have come to associate with quality beef. The trouble is that cows are not designed by nature to eat corn, so they require a cornucopia of drugs to maintain their health.

There's another reason for going into the grass-fed business: it is more consistently profitable than regular beef. That's because ranchers can directly market their beef to local customers, thus commanding premium prices in health-conscious towns such as Durango. It also provides a direct link between the consumer and the producer, a link that puts a human face on eating and agriculture.

For David and Kay this link is crucial: it builds the bonds of community that hold things together.

"When local people are supporting local agriculture," said David, "you know you're doing something right."

Every landscape is unique, and every ranch is different, so drawing lessons is a tricky business, but the lesson of the James Ranch seems clear: traditions can be strengthened by a willingness to try new ideas.

Later, while thumbing through a small stack of information given to me by David and Kay, I found a quote that seemed to sum up not only their philosophy, but also that of the New Ranch movement in general and the optimism it embodies. It came from a wall in an old church in Essex, England:

> *A vision without a task*
> *Is but a dream.*
> *A task without a vision*
> *Is drudgery.*
> *A vision and a task*
> *Is the hope of the world.*

The Allen Ranch
South of Hotchkiss, Colorado

Stand on the back porch of Steve and Rachel Allen's home, located on the western edge of Fruitland Mesa, and you will be rewarded with a view of Stegnerian proportions: Grand Mesa and the Hotchkiss Valley on the left, the rugged summits of the Ragged Mountains in the center, and on the right the purple lofts of the West Elks, a federally designated wilderness where Steve conducts his day job.

I met Steve three years ago when I asked him to speak at a livestock herding workshop I organized at Ghost Ranch, in northern New Mexico. I knew that his grazing association, called the West Elk Pool, had recently won a national award from the Forest Service for its innovative management of cattle in the West Elks. The local Forest Service range conservationist, Dave Bradford, had won a similar award for his role in the West Elk "experiment." Intrigued, I invited them both down to speak about their success.

Steve began their presentation that day with a story. Driving to the workshop, he said, he and Dave found themselves stuck behind a slow-moving truck on a narrow, winding road. At first they waited patiently for a safe opportunity to pass, but none came. Then they grew impatient. Finally, they took a chance. Crossing double yellow lines, they hit the accelerator and prayed. They made it. There was nothing but open road ahead of them, Steve said.

It was meant as a metaphor—describing Steve's experience as a rancher and Dave's experience with the Forest Service. The slow-moving obstacle, of course, was tradition.

In the mid-1990s, Steve and Dave persuaded their respective peers to give herding a chance in the West Elks. They proposed that six neighboring ranchers combine their separate cattle herds into one big herd and move them through the mountains in a slow, one-way arc. By allowing cattle to behave as the ungulates that they are—constantly on the move—the plants are given enough time to grow before being bitten again, which in the case of the cattle of the West Elk Pool wouldn't be until the following summer.

This is unusual because tradition urges ranchers to spread their cattle out over a landscape, especially in times of drought, not bunch them up. Less management is the norm, which is why herding—real herding—is a rare activity in the West.

Dave and Steve described the herding process as a large flowing mass, with a head, a body, and a tail, in almost continuous motion. Pool riders don't push the whole herd at once; instead they guide the head, or the cattle that like to lead, into areas that are scheduled for grazing. The body follows, leaving only the stragglers—those animals that always seem to like to stay in a pasture—to be pushed along.

It is not as hard as it looks, they said at the workshop. When they first began the "experiment" they thought they would need twenty riders to do the job. That just created chaos. Today, for key moves they need only six people, supplemented with the energetic assistance of border collies, and most days the 800-head herd is tended by a solitary rider.

The single-herd approach allows the permittees to concentrate their energies on all of their cattle at once, they said, as well as allowing them and the Forest Service to more easily monitor conditions on the ground.

In fact, the monitoring data showed such an improvement in the health of the land over time that the West Elk Pool recently asked for, and was granted, an increase in their permitted cattle numbers from the Forest Service—this at a time when cattle numbers on public land are headed mostly in the other direction.

Of course, it was more than just the good practice of herding. Once again, it was about a vision. After the workshop, Dave sent me the three-part goal statement for the West Elk allotment.

Our goal is to maintain a safe, secure rural community with economic, social, and biological diversity ... that respects individual freedom and values education, and that encourages cooperation. ... Our goal is to have a good water cycle by having close plant spacing, a covered soil surface, and arable soils; have a fast mineral cycle using soil nutrients effectively; have an energy flow that maximizes the amount of sunlight converted to plant growth and values the seclusion and natural aesthetics of the area.

In an era of increasingly volatile debate about the role of livestock on public land, what Steve, Dave, and the rest of the West Elk Pool have accomplished is quietly significant.

Standing on the Allens' back porch, I asked Steve the question that had been on my mind since the workshop: what set him up for crossing double yellow lines like that?

Steve grew up in Denver, he said, where his father was an insurance salesman. He met Rachel at Western State College in Gunnison, where they discovered that they both liked to ski—a lot. Steve joined the ski patrol in Crested Butte and eventually both of them became ski instructors. It was 1968. They were young and living the easy life. But restlessness gnawed at Steve.

"The ski industry is designed to make ski bums, not professionals," he said, with his easy smile. "It was fun, but we wanted more."

They were also restless about the changes happening in Crested Butte. Even in those early days, signs of gentrification were visible everywhere in town—a sign of things to come.

"The problem with resort communities," said Steve, "is that they attract people who act like they're on vacation the whole year."

Steve and Rachel decided to join the back-to-the-land movement, trading their skies for farm overalls. "We weren't hippies, mind you," interjected Rachel, laughing. "We took farming seriously. I just want to get that on record."

They moved over the mountain to the sleepy village of Paonia, where they planted what eventually became a large garden. They grew vegetables, raised chickens, produced hay, and learned from their farm neighbors.

"Because we admitted we didn't know very much," said Steve, "and because we were willing to learn, people were willing to teach."

In 1977, restlessness struck again. They traded the truck garden for a run-down farm on the edge of Fruitland Mesa, where the hay was so bad the first few years that they had to give it away. They bought a few cattle, and eventually ranching caught Steve's eye. In 1988, he bought a Forest Service permit in the nearby West Elks, mostly as drought relief for his animals. His interest was not purely economic, however. Steve had always been attracted to mountains, and now he had a chance to work in them daily.

Still eager to learn, Steve took a holistic resource management course the same year that Dave James did. It made him restless to give herding a try in the wilderness. With the arrival of Dave Bradford at the Forest Service office in Paonia a short while later, the opportunity to cross the yellow lines suddenly presented itself.

As part of the process of pulling the West Elk "experiment" together, Steve became a student of a new method of low-stress livestock handling sometimes called the "Bud Williams school," after its Canadian founder. Its principles fly in the face of traditional methods of cattle handling, which are full of whooping, prodding, pushing, and cursing.

Instead, Bud Williams demanded that cattle be treated "with respect." All actions should focus on taking stress *off* the animals, not piling it on. There should be no yelling, no electric shocks, and little contact between human and cow. The animals are moved not by coercion but by gentle suggestion, using nature's model of predator-prey relationships. By stepping carefully into an animal's instinctive "flight zone," a human "predator" can gently guide the animal in the direction he or she wants it to go.

John Wayne would not have been pleased. Putting stress on cattle is as customary to ranching as a lasso and spurs. But that was Steve's point: customary, yes; natural, no. And that's where herding comes in: pressure from predators in the wild made ungulates bunch in herds naturally. Unfortunately, on many ranches today, the herding instinct has been prodded out of most cattle.

"Nature," Steve said simply, "has the right ideas, but we keep messing them up."

It is this return to nature's original model, such as grass-fed livestock and low-stress herding, that defines the progressive ranching movement under way today.

Ranching needs good students, but it needs good teachers too.

Grass may lie patiently for the benediction of rain, but people need inspiration. Today, we need it more than ever.

Twin Creek Ranch
South of Lander, Wyoming

Driving through the dry heart of the drought that summer, which was centered in southern Wyoming, I knew the precise moment when I had arrived at Tony and Andrea Malmberg's Twin Creek Ranch. Rounding a big bend in the road, I was suddenly confronted with the sight of green grass, tall willows, sedges, rushes, and flowing water.

Driving up the lush creek toward the ranch headquarters, I recalled a story Tony had written in the Quivira Coalition's newsletter, in an article titled "Ranching for Biodiversity." In it he described an experience from his youth when he and a brother-in-law decided to blow up a beaver dam on the creek:

> *Jim and I crawled through the meadow grass under his pickup giggling. Jim pulled the wires in behind him, leading to the charge of dynamite.*
> *"This will show that little bastard," I said. Jim touched the two wires to the battery. WOOMPH! The concussion preceded the ex-*

plosion. Sticks and mud came raining down on the pickup. As soon as it stopped hailing willows and mud, we scrambled out from under our shield.

"Yeah!" I hollered as we ran down the creek bank. "I think we got it all."

Water gushed through the gutted beaver dam and we could see the level dropping quickly. The next morning I rode my wrangle horse across the restored crossing. The area behind the beaver dam had gotten so deep I couldn't bring the horses across. But that was taken care of now. I galloped down the creek. The water ran muddy and I couldn't help but notice creek banks caving into the stream.

I wondered.

Tony kept wondering, especially after his family was forced into bankruptcy because of high interest payments and tumbling cattle prices, costing them the 33,000-acre ranch.

Two years later, he leased back the ranch before eventually buying it. But he knew things had to be different this time.

Upon completing a course with Kirk Gadzia, Tony began to see that biodiversity was a plus on his ranch, not a minus. "I shifted my thought process to live with the beaver and their dams," he wrote in his article. "With this commitment, I viewed the creek as a fence rather than something I could cross. This attitude gave me an extra pasture, a higher water table, less erosion, and more grass in the riparian area. The positive results energized me, and I began to curiously watch in a new way."

As the beaver returned, he began to notice increased biodiversity. Soon, a University of Wyoming study found a 50 percent increase in bird populations. Moose, previously a rare sight on the property, began to appear in larger numbers. He even began to appreciate coyotes and prairie dogs and the role they played in the health of his land. All of which led him to formulate two guiding principles:

First, I avoid actively killing anything, and notice what is there. Whether a weed or an animal, it would not be here if its habitat were not. I plan the timing, intensity, and frequency of tools (grazing, rest, fire, animal impact, technology and living organ-

isms) to move community dynamics to a level of higher diversity and complexity.

Second, I ask myself what is missing. Problems are not due to the presence of a species but rather the absence of a species. The absence of moose meant willows were missing, which meant beaver were missing and the chain continues.

If I honor my rule of not suppressing life, I will see beyond symptoms to address problems. If I continue asking "What is missing?" I will continue to see beyond simple systems and realize the whole. When I increase biodiversity I improve land health, I improve community relations, and I improve our ranch profitability.

To accomplish his goals, Tony employs livestock grazing as a land management tool. To encourage the growth of willows along the stream and ponds, for example, he grazes them in early spring to aid seedling establishment. By concentrating cattle for short periods of time, Tony breaks up topsoils and makes the land more receptive to natural reseeding and able to hold more water.

What brought me to Twin Creek, however, wasn't the tall grass, the flowing water, or the progressive ranch management exhibited by Tony and Andrea, though these were important. What I wanted to see was the very nice bed-and-breakfast they operated.

As I pulled up to the spiffy, new, three-story lodge, I was greeted with a sunny wave by Andrea. A child of the Wyoming ranching establishment—her father traded cattle for a living—Andrea heard Tony speak passionately about the benefits of planned grazing at a livestock meeting some years ago (where his talk was coolly received) and wrote him an equally passionate letter challenging his beliefs. They corresponded back and forth until she accepted his dare to come to the ranch and see the proof herself.

Later, over a glass of wine, I learned that the lodge had its roots in anger. "When my family lost the ranch," recalled Tony, "I blamed everyone but ourselves. I blamed consumers, environmentalists, liberals. But most of all, I blamed our new neighbors."

In 1982, as the family was slipping into bankruptcy, a man from

California bought a neighboring ranch for twice what a cow would generate per acre.

By the time Tony returned to the property, however, he had a revelation: markets don't lie. In addition to the cattle operation, he knew it was time to start a ranch-recreation business and market it directly to people such as his new neighbors. He learned quickly, however, that paying guests didn't like dirt or mice as much as he did, so he and Andrea took the plunge and built a pretty lodge with a capacity for fourteen guests at a time.

But they didn't stop there.

Andrea convinced Tony that the next step was to "go local"—meaning, find ways to tap local markets. They started by hosting a class on weed control for new, local ranchette owners, focused on goats, which will eat every noxious weed on the state list.

It was a big hit.

That was followed by a seminar on rangeland health, which proved popular with their ranching friends. Then came a foray into the grass-fed beef business (which has been successful too).

Next up was Andrea's decision to start teaching yoga lessons. A recent winter solstice party packed the lodge with what Tony admitted was the "strangest assortment of people I'd ever seen together."

Yet there was a lesson here too.

"The hodge-podge appeared to be a demographic accident," he continued, "yet they all ended up in central Wyoming because they wanted the same things we want: a beautiful landscape, healthy ecology, wholesome food, and a sense of community."

In this, Tony drew a parallel to the benefit of increased biodiversity on the ranch.

"In the old days, I didn't have to deal with people different from me," he said. "But this is better."

Tony explained how his indicators of success have changed over the years, reflecting not only their personal journey but the New Ranch movement in general.

In 1982, his primary measure of success was a traditional one: increased weaning weights of his calves. By 1995, Tony's measure had shifted to stocking rate (of cattle), which was up 75 percent from years prior because of their planned grazing. By 1998, the indicator shifted to monitoring—transects of land health—and what the data said about trend, up or down (up mostly). By 2000, Tony used the diversity of songbirds on the property as his baseline (more than sixty species). By 2002, however, the main measure of success had changed to how many values generated income for the ranch in a year (three).

Tony explained the last shift this way: "If we're going to manage for biodiversity we need to be able to get paid for it. And we need to educate people to be responsible and accountable for how they spend their dollars."

And in order to do the education properly, they have learned to speak various languages.

"I realized that if I'm going to survive in the twenty-first century, I need to be trilingual," Tony explained. "Ranchers tell stories. The BLM wants to talk data. And then we've got the environmentalists. Lander has a lot of them. To connect with them you need to use poetry."

In other words, success in ranching today is as much about communication and marketing as it is about on-the-ground results. As the farm population across the United States continues to dwindle—less than 2 percent nationally, according to the U.S. Census—it has become imperative that farmers and ranchers find new ways to be heard, both economically and politically.

As Tony's story suggests, it is not enough to simply *do* a better job environmentally, even if it brings profitability. One must also *sell* one's good work, and do so aggressively in a social climate of rapid change and increasing detachment from our agricultural roots.

From all the indicators I saw, Tony and Andrea are on the right track. The lodge was clean, comfortable, and airy; the food wonderful; and the visitors happy.

But this is no dude ranch. Tony makes his guests work; in fact, he very cleverly employs them in his progressive ranch management. According to his planned grazing schedule, his cattle need to be moved almost every day—so he has paying guests do it. They love it, of course, and since his cowboy does the supervisory work, Tony is free to explore other business ideas.

And the ideas keep coming.

Red Canyon Ranch
West of Lander, Wyoming

When I met Bob Budd at The Nature Conservancy office in Lander, he was pacing the floor, waiting for my arrival.

"The ranch is on fire," he said quickly. "Let's go."

And go we did. Despite being a foot taller than Bob, I had to hustle to keep up with him as we headed outside. A Wyoming native son, member of a well-known ranching family, and former executive director of the Wyoming Stockgrowers' Association, Bob managed the Red Canyon Ranch for The Nature Conservancy's Wyoming office when I met him. He also served as their director of science. If that wasn't enough to do, Bob earned a master's degree in ecology and was in line to become president of the Society for Range Management, a highly respected international association of range professionals.

Without a doubt, he was a man on the move.

I jumped into my truck and followed Bob rapidly to the headquarters of the Red Canyon Ranch, which borders Lander on the south and west. The Nature Conservancy, Bob said when we arrived, had purchased the property for three reasons: to protect open space and the biological resources held there, to demonstrate that livestock production and conservation are compatible, and to work at landscape-level goals.

The first two goals have been achieved, more or less, he said, as I jumped into his truck. It is the third goal that motivated him now. What Bob wants is fire back on the land, brush and trees thinned, erosion repaired, noxious weeds eradicated, perennial streams to flow more fully, riparian vegetation to grow stronger, and wildlife populations to bloom.

And judging by the speed at which we traveled, he wanted them all at once.

Bob was thrilled about the lightning-sparked fire, for example, which was burning a chunk of forest and rangeland right where he had been encouraging the Forest Service to light a prescribed fire for years.

"I love lightning," he said with a twinkle in his light blue eyes, "because there's no paperwork."

As we sped into the mountains in search of a suitable vantage point to observe the progress of the fire, talking energetically about the ecological theories of disturbance, homogeneity vs. heterogeneity, cascade effect, scale, and so forth, I recalled another accomplishment in Bob's long list of achievements: that of essayist.

Three of his essays, in fact, had been published recently in a book titled *Ranching West of the 100 Meridian*. In one he wrote: "I am an advocate for wild creatures, rare plants, arrays of native vegetation, clean water, fish, stewardship of natural resources, and learning. I believe these things are compatible with ranching, sometimes lost without ranching. Some people call me a cowboy. A lot of good cowboys call me an environmentalist."

Bob has strong words for both, especially about their respective defense of myth. He likes to remind environmentalists in particular that nature is not pristine, as many assume. For thousands of years, Wyoming has been grazed, burned, rested, desiccated, and flooded. In saying so, he consciously tilts at an ecological holy grail called the "balance of nature."

"In landscapes where the single ecological truth is chaos and dynamic change," he wrote, "we seem obsessed with stability. Instead of relishing dynamic irregularities in nature, we absorb confusion and chaos into our own lives, then demand that natural systems be stable. We ask systems that evolved in geological time to respond and perform in our own lifetime."

He likes to tell both environmentalists and ranchers that grazing, like fire, is a keystone process in North American ecosystems.

"Like fire, erosion, and drought, grazing is a natural process that can be stark and ugly," he wrote. "And, like fire, erosion, and drought, grazing is essential to the maintenance of many natural systems in the West. ... And because adults tend to overlook other grazing creatures, we forget the impact of grasshoppers, rodents, birds, and other organisms that have long shaped the West."

Bob observed that prescribed fire, once controversial, is now widely accepted. He thinks it is simply a matter of time before the same change of thinking happens to grazing. In the meantime, he worries about scale. Humans tend to think and work at human-sized scales, in both time and place, but nature works far differently.

"If we wish to maintain intact systems," he concluded in his essay, "we must learn to manage and inquire on a scale that recognizes biological lines rather than lines of property ownership. What is best for the landscape will only be realized when it can be accomplished in a manner that follows the flight of the falcon. ... And intact systems are ranches."

I asked him about this observation as we sped through the forest, still searching for a spot to view the fire.

"Western ranches are often disparaged for damage they do to natural systems," he said, "yet much of the natural landscapes and crucial habitats in the West are directly tied to the future of the ranching industry. In Wyoming, as throughout much of the West today, unbridled development has resulted in habitat fragmentation and destruction. As land is subdivided, associated roads and human development often interrupt wildlife migration corridors, decrease habitat for rare plants and animals, and make ecosystem management ever more difficult. Ranchlands are the final barrier to this type of development in many areas. The economic viability of ranching is, therefore, essential in maintaining Wyoming's open space, native species, and healthy ecosystems."

As we traveled, Bob pointed out the unhealthy condition of the trees we passed through and listed restoration actions that are required to get the forest back into shape.

"Our common goal must be to provide the full range of values and habitat types that a variety of species need, including us," he said. "We achieve it by assuring that there are sites in various states of degradation, maturation, and successional movement both toward and away from the extreme. We need to be thinking in terms of the full range of variability on a landscape if we want to manage for biodiversity and long-term sustainability."

In other words, ranchers need to become restorationists. In fact, ranchers are, in Bob's view, uniquely positioned to deliver ecological services—as landowners, as livestock specialists, and as hardworking sons- (and daughters)-of-guns.

Suddenly, we stopped. The fire had proved elusive, and it was time to turn around and head back to headquarters. I couldn't ignore the symbolism. While landscape-scale opportunities for ranchers may be plentiful, as Bob suggested, so are the obstacles, especially on public land, where every action seems to engender an equal and opposite reaction by someone. Even the smallest restoration project, whether it involves livestock or not, can very quickly become mired in red tape and conflict.

Nevertheless, Bob said, he remained optimistic. He admitted that he had to be.

Returning to the ranch headquarters, Bob kept moving. He needed to take his son to baseball practice. I followed him into the house for introductions to the family. We talked for a while longer, shook hands, and before I knew it, Bob was gone.

Rather than drive off immediately myself, however, I walked down to a bridge that spanned the burbling, contented creek. Enjoying a momentary respite from the dust, driving, and cascade of ideas, goals, and practices that dominated the trip, I leaned against a wooden post and listened to the wind.

One thread that tied Bob Budd to the Jameses, Allens, and Malmbergs, it occurred to me, was the desire to heal. Whether it was restoring land to health, strengthening community relations, teaching, feeding, or peacemaking, every person I encountered was engaged in an act of healing. This is good news for grass, especially in these dry times.

Perhaps ranching *is* immortal, I thought, just not in its current form. After all, humans have been living and working with livestock for a very long time and through a great deal of change. The need to be near animals, and be outdoors, hasn't altered much over the centuries. In the Next West, how it gets expressed will just be different, again.

Drum Hadley has been a rancher and a poet along the southwest border-lands for over four decades. He is cofounder of the Malpai Borderlands Group and originated and implemented the first grassbank. His most recent book of poems is Voice of the Borderlands. *In this poem, he describes a cattle drive in Sonora, Mexico.*

Who Are We Here, Wanting to Know?

Alma de mi alma
A Song of the Vaqueros of Mexico

Drum Hadley

El sauce y la palma se mesclan con calma.
Alma de mi alma que linda eres tu.
'The willow and the palm,
They gently touch each other.
Alma of my soul, soul of my soul,
How beautiful you were.

Alma born 1953 to Don Cruzita Alonzo,
Vaquero in the Cañon de Dimas,
Where the swallows come nesting,
By the red cliffs in the Springtime Sonora, Mexico

Where do the swallows go, passing with the west wind?
By the red cliffs they stay for a while, then go.
When she was a young girl, her mother didn't want her,
So took her to town to Doña Petra.

Where do the swallows go, passing with these west winds?
Where do they nest for a while and then go?

When she was sweet sixteen, Doña Petra didn't want her,
So she went to live with her Uncle Peru.

She was shot through the heart by Peru's jealous wife.
She's buried in the *cañon* at El Ranchito.
Where do your hates and your jealous loves go?
Who are we here ... wanting to know?

Who are we here, wanting to know?

Shy, whirling Alma, dancing your young-old eyes.
The carousing *vaqueros* chased all night,
Till the sunlight lit their camp on the town street
Between our roundup *jefe's* house and hers.

Como un águila bajando a un lepe,
Roberto bailó con las señoritas in Agua Prieta
Like an eagle, dropping down on a dogie calf,
Roberto danced with the señoritas in Agua Prieta.

Where will the old Earth take you, dancing through the starlight?
Whirling you on and on, while she goes.
Where will the old Earth carry us, dancing through the starlight?
Whirling us all on and on while she goes.

Danced her through that old white house, where she lived.
One room, adobe mud, the other of cardboard and rusting pieces
    of rattling tin,
Where Petra served us *frijole* beans and *carne,*
As though we'd come driving steers along the dusty trails as kings.

Whirling you on and on while she goes,
Whirling us all on and on while she goes,
West from the San Bernardino River, through Gallirdo Pass we rode
In the dusk light, lost two steers in the night-time.

Rode on again another day into Agua Prieta,
Through dirt streets and Mexican kids
Running by the sides of the road, throwing rocks at stray steers,
To keep them headed towards the border corrals.

Then with tequila and *corridos* floating through the cantinas,
And the women, and the songs, we forgot the dust, the wild cattle,
The cold of the mornings, the winding trails, and changed the town
To some whirling place, we didn't know or remember.

Where do those nights and the singing in your memories,
And the crossings of these valleys, and the sandy rivers go?
Where do those nights and the singing in our memories,
And the crossings of these lands, and the sandy rivers go?

And the *vaqueros*, who rode whistling
In those soft dark eyes,
While the swallows circled and drifted in the winds
Calling by the red cliffs in the Cañon de Dimas in Springtime.

Where will those loves, and your laughing black eyes,
And the winding river go?
Who are we here, wanting to know?
Who are we here, wanting to know?

—*Voice of Carlos Yslava*

Rick Bass is the author of twenty-two books, and his stories have been awarded the Pushcart Prize and the O. Henry Award. Though he is best known for his relationship with and advocacy for the Yaak Valley in upper Montana, he here turns his attention to his birthplace—Texas. In "The Farm," he gives us a story of his mother, his children, and the link to a place that binds them all.

The Farm

Rick Bass

It was still the end-of-winter at our home in northern Montana, but down in south Texas, in April, at my father's farm, it was full-bore spring. It was a joy to me to realize that Lowry, just-turned-three, would now have the colors and sights of this place lodged in at least her subconscious, and that Mary Katherine, just-turned-six, was old enough to begin doing some serious remembering. Some children of course hold on to odd-shaped bits and pieces of memory from a much earlier age—but around the age of six and seven, nearly everything can be retained—or at least that was how it worked for me, when I was a child.

It was like a kind of freedom—a kind of second welcoming her into the world. Now when I was an old man I would be able to say to her, "Remember when ...," and she would remember.

We had flown to Austin, rented a car, visited my brother, and then had driven down into the brush country and toward the live oaks and dunes that lay in braided twists some fifty miles inland, to the farm. As we drove, Elizabeth and I talked and watched the late-day sunlight stretch across the green fields; the girls slept, tired from their travels, in the back seat. Angels. So much joy do they bring me that sometimes I wonder if, since my mother is not here to love and know them, I'll carry also her share, having inherited prematurely her portion of that joy. For

124

a fact, this joy seems too large. I think maybe that is what is happening, sometimes, at certain moments. I glance at them, and love them fully and deeply, but then a second wave or wash comes in over that one, as if she is watching them over my shoulder, and I feel it again.

It used to give me a bittersweet feeling, but now I'm not sure what the word for it is. Gratitude, sometimes: to the girls, of course, but also to my mother.

They woke when we stopped to open the gate. We drove through and closed the gate behind us, and because we could not wait, we parked the car there and decided to walk instead of drive the rest of the way to the farmhouse. We walked in the late-day light, the last light, down the white sandy winding road, beneath the moss-hung limbs of the enormous live oaks—trees that were five and six hundred years old. It's so strange, the way there will be certain stretches of time, certain moments, where for a little while it will feel exactly as if I am walking in her every footstep: as if I *am* her, in that moment, set back in time—and enjoying that moment as I know she must have enjoyed it, or one like it, thirty or forty years ago. And I wonder, is it just this way for me, or do others experience such feelings, such moments?

Buttercups, winecups, and black-eyed Susans; before we had taken ten steps, Lowry and Mary Katherine both had picked double-fistful bouquets, and had braided flowers in their hair. Another ten steps took us across the culvert that ran beneath the road. There was water standing in the culvert and in receding little oxbows on either side of the road, and as we approached, ten thousand little frogs went splashing into that muddy water. "Frog alert, frog alert!" we cried, and ran down to mud's edge to try to catch one, but there were too many, springing zigzag in too many directions; you couldn't focus, and couldn't chase just one, because their paths were crisscrossing so. There were so many frogs in the air at any one time that occasionally they would have midair collisions; they were ricocheting off each other. The mud around the shoreline of their fast-disappearing pond glistened, so fast was the water evaporating, and the mud was hieroglyphed with the handprints of what might have been armies of raccoons, though also it could have been the maddened pacings of one very unsuccessful raccoon.

We finally caught one of the little frogs, and examined it: the gray-brown back that was so much the color of the mud, and the pearl-white underbelly. I wondered why, when frogs sunned themselves, they didn't stretch out and lie on their backs, the way humans do at the beach. I guess they would get eaten. I guess if a frog had a mud-brown belly it could lie on its back, camouflaged to the birds above, and still be able to listen for the approach of terrestrial predators, but I guess also there's no real evolutionary advantage to a frog being able to warm its belly in the sun. Though for that matter the same could be said of us.

Into the farmhouse she loved so much—she had lived in it, and loved it, for only a few years before she fell ill, but had loved it so fully in that time that I still cannot step into it without feeling that remnant love-of-place. And it is thin substitute for her absence, but with the exception of my own blood in my veins, and memories, it is all there is, and I am grateful for it, *place.*

Elizabeth wanted to go for a run in the last wedge of light—after the long Montana winter we were nearly delirious with the gift of these longer days—and so she laced up her running shoes and went on back up the road at a trot. Mary Katherine wanted to go fishing in the stock tank, so we rigged up a line and went off toward the pond, following the winding sand road and walking beneath those old trees.

We stood on the levee and cast out at the ring of flat water. Turtle heads appeared in the center of the lake, tipped like little sticks, to observe us. In the clear water of the shallows we could see the giant Chinese grass carp, thirty pounds each and seemingly as large as horses striding just beneath the surface, cruising; my parents had put them there when they first built the pond as a means of keeping algae from overtaking the pond. The carp are hybrids, so that they can't reproduce, though it's rumored they can live to be a hundred years old—and because the carp are strictly vegetarians, there was no chance of them striking at our spinnerbait. It was strange, though, watching the giant fish circle the pond so slowly, their dorsal fins sometimes cresting the surface like sharks, and knowing that we were fishing for something else, something deeper in the pond—fishing for fish-beneath-fish.

On the far side of the pond, a big fish leapt—not a carp, but a bass. We cast to it for a while, in the gathering dusk, but I was hoping that we wouldn't catch it. It's good for the girls to learn that you don't get something every time you go out, or right away.

A water moccasin swam past, its beautifully ugly wedge of head so alarming to our instincts that it seemed almost like a mild form of hypnosis—as did the eerie, elegant, S-wake of its thick body moving across the surface. There were floating four-leaf clovers that my parents had planted—a special variety in which every one of them had four leaves—and we stopped fishing for a moment and picked some for friends.

Across the field, across the rise, we could see the cattle trotting in front of the blood-red sun, running from something, and in that wavering red light, and across the copper-fading visage of the pasture, it looked like some scene from Africa: some vast herd of wildebeests. The cattle passed from view and then a few moments later we saw the silhouette of Elizabeth jogging along the crest of the rise—she was what had spooked the cattle—and across the distance we watched her run in that Mars-red light, the sun behind her, as seven years ago I had sat by this same pond with my mother and watched Elizabeth and my father ride horses across the face of that sun.

We resumed casting. A mockingbird flew up and landed in the little weesatche tree next to us, not five hundred yards away, and as the sun's fireball sank (as if into an ocean), the mockingbird began singing: some intricate melody which, in the blueing of dusk, and then the true darkness, was one of the most beautiful songs—a serenade—I've ever heard.

"Sing back to him," I told Lowry, and so she did; she sang her alphabet song there in the darkness, her "A-B-C's—next time you can sing with me"—etc.

Finally it was true dark—the mockingbird was still singing—and we headed back toward the house. We saw a shuffling little object, a humped little creature, shambling down the sand road in front of us, and I cried "Armadillo! Chase him!"

We set out after him in full sprint, and were almost even with him— he was running in zags and weaves through the trees—when I noticed the white stripe running down his back and was able barely in time to shout, "Skunk! Get back!"

Perhaps it was the four-leaf clovers. The skunk went his way, and we went ours. I had the strangest thought, in my relief, however. I found myself wondering how—had we been sprayed—the girls would have thought of me afterward, growing up. What if they grew up to be storytellers? What kind of mirth would they have had with that—recounting, for the rest of their adult days, the time their father told them to chase and catch a skunk?

How lucky they were, by fluke chance, to remain in normalcy, and to escape unsprayed, untraumatized; and how lucky I was, by the matter of a few feet, to not have such identity fastened to me by my children, with the permanence of myth.

I remembered the time when I was about Mary Katherine's age, when my cousin Randy was sprayed by a skunk. It was right around Christmas. We were all gathered up at Grandma and Granddaddy Bass's, in Fort Worth—my parents, brothers, and myself; Aunt Lee, Uncle Jimmy, and my cousins—Rick, Randy, and Russell. I had already gone to bed—I think it was Christmas Eve—but Randy, being a few years older, was allowed to go down to the creek to check his trotlines and his Hav-a-Hart trap one more time.

I had just nodded off to sleep when I awoke to the impression that all the doors in the house had been blown wide open by some awful force. All of the adults had just let out a collective roar—a primal group groan—and then there were gasps and more groans, and my uncle's voice, angry and above all others, "Randy, get *out* of the house!"

Then the smell hit me. Even in the back room, it was stout. I hadn't known that an odor could be that powerful. It seemed that it could levitate the house. It certainly levitated the people in the house.

When I went out to ask what all was going on, I seem to recall a furious, sputtering inarticulation on the part of the grown-ups, until finally—or this is how I remember it—they shouted, in unison and choreographed with much arm-waving, *"Randy!"*—as if that said it all.

Thirty-plus years later, the girls and I let the skunk travel on his way, and we went ours, still sweet-smelling. We could see the glow of the farm-house through the woods and were striking toward it, holding hands and walking carefully in the darkness to avoid stepping on any skunks, when I saw a firefly blink once, then twice, in the distance, and I shouted with happiness.

The girls had never seen fireflies before. I am not sure they had even known such creatures existed.

For the next hour, we chased them through the meadow, trying to catch just one. It seemed a harder task than I remembered from my own childhood— I remembered filling entire lantern-bottles with them— and I figured that it might be because it was still early in the spring and they were not yet blinking with full authority or intensity. We'd see only an individual blinking, and always at some great distance. We'd break into a run, hoping to arrive there before the blink faded, but they were always a little too far away, and their luminescence lasted only a few seconds. We would leap at that last instant, toward the always-ascending (they heard us coming) fading glow of gold—leaping with cupped hands and blind faith toward some imagined, calculated place ahead of us where we believed their flight path would take them—and opening our hands cautiously then, in the silver moonlight, to see if, like a miracle—like plucking a star from the sky—we had succeeded in blind-snaring one.

As beautiful as the on-again, off-again drifting missives of the fire-flies was the seamlessness with which Mary Katherine accepted unquestioningly the marvel of such an existence, such a phenomenon. As if secure almost to the point of nonchalance, or at least pure or unexamined wonder, that Yes, of course, this was the way all silver-moon nights were meant to be passed, running and laughing and leaping with great earnestness for drifting, blinking low-stars against a background of fixed, higher stars.

Eventually, we caught one. And one was enough. We went through the time-honored ritual of putting it in a glass jar and punching air holes in the top. We took it inside the house: turned off all the lights. That simple, phenomenal, marvelous miracle—so easy to behold—even as old familiar things have left us, replaced by a newness in the world.

The heck with electricity, or flashlights. *Yes. This is the world my daughters deserve. This is the right world for them.*

Later that night, after a supper cooked out on the grill, and after the girls were asleep (dreaming, I hope, of leaping), Elizabeth and I went for a long walk in the moonlight. The brightest, most severe, platinum light I have ever seen. Revealing more, in the glare of that intense silver-blue light—highlighting certain things—than would the normal broad light of day.

It didn't feel as if we'd been together nearly twenty years. Or rather, part of it did: the good part.

Such strange brilliance.

The next day we all went fishing. It was windy, and Elizabeth's straw sunhat blew off and landed rightside-up on the pond. We watched it sail quickly, without sinking, all the way across the little lake. Mary Katherine ran around to the other side of the lake and was there to fish it out with a stick when it arrived. She ran it back to Elizabeth, who put it back on and tied it tighter this time.

The joy of children catching fish: there's nothing like it. Most of the few fish we were catching were too little, and we kept throwing them back. Low's pink skin, her bright blonde hair, in that beautiful spring sun. A hundred feet of snow, it seemed, back home, where we live now, though this was once my home, south Texas, and it's a wonderful feeling to be able to come back to a place that you have left and to feel that place welcome you back, and to feel your affections for it undiminished across time.

Mary Katherine kept wanting to keep some fish for dinner that night. We finally caught one that was eating-size, and as I put the fish on the stringer, I said "It's your unlucky day, my little friend." And for the rest of the day, whenever we'd catch one, Mary Katherine (while hopping up and down and clapping her hands, if I'd caught it, or simply hopping up and down, if she'd caught it) would say, "Oh, please let it be his unlucky day, oh please!"

That night, after our fish fry, I took her into town for an ice cream cone; Lowry had already fallen asleep. I get so used to doing things with them together, the two girls, that I have to remember this: to always be there to spend some time alone with each of them. The special quality or nature of that as unique, as sacred, as the quality of sunlight early in the spring, seen dappled through a new green canopy of emerging leaves, or in the late fall, when the light lays down soft and long again after the harsh bright summer.

It was dusk again, and nighthawks were huddled along the edges of the white sand roads as we drove slowly, twisting and turning, beneath the arched limbs of more old live oaks. Fireflies were out in the meadows again and we rode with the windows down to feel the cool night air. The radio was playing very quietly—a jazz special, with the music of Sonny Rollins and Louie Armstrong—and I knew by the way Mary Katherine rode silently, happily, that she had never heard such music before.

We got into the little town of Yorktown a few minutes before closing time. We went into the coolness of the air-conditioned Dairy Queen and I waited while Mary Katherine pondered her selection, deciding finally on a chocolate dip cone.

She ate on it the whole way home—back through the starry night, back through the fireflies, back to the rest of our waiting, sleeping family—just riding and listening, all the while, to that strange, happy, lulling music from so long ago.

I have never felt more like a father: never more in love with the world.

Sharon Salisbury O'Toole and her husband, Patrick, raise cattle, sheep, dogs, and children on their five-generation family ranch near Savery, Wyoming. They probably trail livestock more than any other transhumance operation in the West and are greatly impacted by the oil and gas development in the region. "Atlantic Rim: The Seekers' Trail" speaks to this impact.

Atlantic Rim: The Seekers' Trail

Sharon Salisbury O'Toole

Earth's creatures tread the ancient trails,
    Dusty paths from grass to grass,
    From summer's green to winter's sage
    On elk-trod road pass deer and cow,
    Hooves of sheep, all those who graze,
    No gas-fired truck to speed their way.

But wells which belch this gas have changed
    This time-worn path, this trodden trail,
    For those who walk and those who graze
    Through winter grass, through browse, through dust.
    Oil-field roads now cut this land
    Of sage and sand. For those who watch,

Who watch from desert heart and sky,
    Raptors feed on road-kill feasts.
    Grouse crouch low beneath the sage,
    Coyote sniffs for scent of prey,
    Rattler shimmies through the dust. All
    Feel the rumble, hear the thrum as

Machinery parts these waves of sage.
    Primordial seas laid lodes of gas
    That heat our homes. We build these roads
    Through hoof prints laid on age-old trails,
    Through bones and seeds and trodden dust
    Raised by those who move and graze.

They graze along with season's change
    Through sky and grass and scent of sage,
    Through cold-bit snow and shining dust
    Back-lit by flaring gas-fueled flames
    Trucks now cross a hundred trails
    And roar on roads of new construct.

Land first scarred by two-track road,
    West-bound wagons whose oxen grazed,
    Drew seekers on the homestead trail,
    Cross sage and stream, cross nation's heart.
    Now gas-field trucks roll on, roll on,
    Dusty contrails riding high.

Sky filled with dusty trails that track
    Where trucks and roads and people go.
    Gas wells squat with painted tanks.
    Behind them ranks of antelope graze.
    Witnessed by the timeless sage,
    And those who tread the ancient trails,

Stock and game on hoof-worn path,
    Winter's bounty led them here
    Joined now by trucks and roads and dust.
    We moil for gas, for coal, for oil,
    This path laid down by those who search
    This sagebrush trail where seekers go.

Paul F. Starrs is interested in the survival of entire landscapes, including their physical features, people, economy, and past life. In "Words Meet Deeds on the Land," he looks at why ranching is a sentimentally regarded feature of the West, and one that would be missed were it to disappear from the scene.

Words Meet Deeds on the Land

Paul F. Starrs

Twenty-five years ago, after a couple of horseback wrecks left me with a shoulder that pops out with the benefit of a good cough, I hung up my spurs, literally: strap ends buckled together, they colonize a nice spot on the wall, hanging as a still-life reminder of the twenty-first-century trappings of an aging "cowhand." Yet I have not given up thinking about the experience of a half-dozen years in which buckarooing a thousand square miles and four hundred agile cows was part of my life.

Eventually, each of us catalogs our aches and dents in the hard work of growing old. Finesse and thought overtake raw strength and physical virtuosity. We age, reflect, judge, and look outward. A lot of memorable landscapes are opened with time spent in ranching. And the slow pace of generations spools by. There is another set of lessons learned in watching our children—the next generation—reach their mid-teens, to start contemplating their options for later in life.

For those already part of a ranching family, the choice to embrace ranching isn't an easy one; for those outside, looking at something deep in tradition and literature, they see pathways to entry that seem awfully slender and sinuous and expensive. If they work off the ranch, it's so they can go back as often as they can. For the people who want to become ranchers, there's hope that a day job can support weekend or

While the traditional signal that a cowboy has retired is "selling his saddle," I don't know any who have: that's the last thing you'd part with, which perhaps is why that act is taken to have such finality. But hanging up your spurs is equally symbolic, though a lot less permanent. The cowhand who has transformed into something else often festoons a residence with evidence of his or her bygone life, so ropes and artwork and hats and elegant but practical bits or hackamores or horsehair reins and mecates loom with significance. (Photograph by Paul F. Starrs, March 2006)

vacation jaunts to someplace where ranching remains alive. The ranch has long been a sought-after goal. The attraction, though, is clear: ranch life, social and complex, has its own poetry.

Plenty of ranchers do what they most want to, and garner a sense of worth from exactly that. Ranching is seeing itself transformed from a commodity-based activity to embrace "amenity-based" reality, as the geographer John Holmes reminds us, which variously emphasizes conservation, tourism, quiet, open space, prestige, wildlife, clean air and water, and the good life. Those advantages are cherished and shared by ranchers in remarkably dispersed grazed corners of the planet.

In looking at ranching, one lesson is clear: it is a better way of life than a road to easy money. Livestock ranching in the developed world is far more social than it is cash-based. A life is encapsulated in ranching that joins time working with land, living close to animals domesticated to suit our needs, aesthetics, and affection, to a now too-little-understood natural world. We find again our affection for landscape. Ranch work is lauded in novels and short stories, song, nonfiction, and photography. And in a wonderful switch, a difficult and spare way of life is praised in—of all things—poems.

Elko-Bound

Not long ago, I gave the keynote address at the National Cowboy Poetry Gathering in Elko, Nevada, at that point in its twenty-second year. Being a semi-big-shot performer was a kind of guilty pleasure, since I'd seen "cowboy poetry" in Elko for the first time in 1985. Tellingly, my predecessor at the year-before "gathering" in 2005 was the indefatigable historian of the New West, Patricia Nelson Limerick, who chose "The Urban-Rural Divide" in the American West as her theme.

My topic was poetry and why it's written about cowboys—and it was a departure not because I don't see the same divide that Patty captured and dissected, but because I see "cowboy poetry"—and especially the dauntless poetry practitioners and songsters who come annually to Elko (and other venues), attempting to bandage, splint, steward, and make whole our relationship to the West's land and its inhabitants—or at least, tell some good yarns and recite some classic cowboy poems. Right there is the interesting divide: are ranchers despoilers, as some would have us think, or stewards? Environmental protection and quality are inescapable themes, and work by a number of scientists increasingly suggests that biodiversity and environmental quality are safeguarded far more by steady and knowledgeable stewardship than by passing fancies of short-term managers from a federal agency or corporate HQ or real estate office: the long view protects.

While opinions span a fairly broad range, I opt for the view of the Jewish cowboy-songwriter-novelist-politician Kinky Friedman. In an editorial for the vaguely venerable grand old lady of national newspapers, the *New York Times,* the Kinkster offered a calculus of the cowhand: "True cowboys must be able to ride beyond time and geography. They must leave us a dream to grow by, a haunting echo of a song, a fine dust that is visible for generations against even a black and white sunset."

Home Ranch. Now part of the Toiyabe Cattle Company, this sparks an interesting bit of argument. I once claimed there was transcendent significance to the name, only to have Maureen Hurley of Oakland, whose family, the Walshes, came to the Reese River Valley in the 1860s, explain that it was called "home ranch" simply because it was their first home in the New World. (Photograph by Paul F. Starrs, October 1994)

Where literature meets an elemental and ancient human endeavor—pastoralism or herding—there is a story to be told. Some, like the explosively original Blake Almendiger, serve up eye-watering visions of the Cowboy West in which the working hand is a hopeless symbol of delusional innocence. Others, including my former student and sometime inspiration Barney Nelson, an original Elko cowboy poet, have offered distinctive and sage analyses of the linkages of a ranch hand with the land, but it is appropriate for us to dwell here upon another toothsome theme that brings its own postmillennial message and fascinations: the craft of the cowboy bard.

The Alienated Cowboy?
And of the cowhand or the rancher, what? That's another matter, but not very far afield. The cowhand—whether buckaroo or Texas waddie,

and in fact, whether gaucho or vaquero or charro—shares a trait the world round (and cowboying is a worldwide practice), and that is an appreciation for humans working in long-standing contact with domesticated animals, and being shaped by a continuous relationship with the physical world. Those "domesticated animals" can include sheep, cattle, routinely horses, and I won't willingly exclude goats—though most hands would turn pale at that. Other countries, other pastoralists, have their own animals, and so long as those are relatively big and relatively rugged, and have been a part of the human bestiary at some point since Noah, they might as well qualify.

The point is, there is a craft to the trade of the animal tender. Honesty commands a concession that some of those trades have more innate romance than others. But the title "keeper of the herd" applies whether the people involved are hired hands or owners, so the life of the hand is very little removed—except by financial responsibility—from the ornate economic sagas of the ranch owner, who if anything would have more to lose in failing to tend the herd. We call this person a "steward" or caretaker or herder, and each of those terms is framed with affection for our charges that will not quit.

No less momentous a personage than the western literary historian, rancher, and University of Texas professor J. Frank Dobie famously commented in his *Guide to Life and Literature of the Southwest* that "the cowboy became the best-known occupational type that America has given to the world. He exists still and will long exist, though much changed from the original." Dobie goes on, "Romance, both genuine and spurious, has obscured the realities of range and trail. The realities themselves have, however, been such that few riders really belonging to the range wished to lead any other existence. ... Bankers, manufacturers, merchants, and mechanics seldom so idealize their own occupations; they work fifty weeks a year to go free the other two."

Any Marxist among us would be inclined to sarcastically remark that sympathetic talk of ranching's "romance" clearly constitutes part of an elaborate plot designed to distract the cowboy from an essential alienation from the means of production. In other words, being a cowboy is so cool that hands are willing to put up with horrible and dangerous work and impossible hours and almost nonexistent pay because

Some of the other domesticated herbivores, a reminder of the close proximity that some humans maintain to the animals whose history with humans dates back some 10,000 years. (Photograph by Paul F. Starrs, Summer 2004)

"being cool" is its own reward. That, the materialist mind would say, keeps the cowboy in his place and prevents rebellion against the ranch owner. But here again Dobie ripostes: "Many scores of autobiographies have been written by range men, perhaps half of them by cowboys who never became owners at all." There is nothing comparable in world literature: stonemasons, aqueduct builders, architects, proctologists, even psychiatrists and lawyers have nothing that begins to approach the vibrancy of the cowboy yarn, whether presented in prose, film, photography, song, cuisine, scent, or poem.

Yes, that is a substantial claim, but not one made lightly. As a scholar, Dobie's Texas affiliations rendered him opaque to some of life's brasher realities, a problem not unknown to the adopted sons of that region. But Dobie was smack-dab on top of it when it came to writing: there are thousands of books and therefore hundreds of thousands (it would not be an exaggeration even to say millions) of pages of literature on the cowboy theme. And some of the very best writing on a cowhand's existence—by the likes of Nevada's Owen Ulph, for example—are in actual fact creations of the 1980s and '90s (though Larry McMurtry, Linda Hasselstrom, J.P.S. Brown, and Mac Hedges are not a whole lot further off, and Cormac McCarthy is even better in some ways, if harder to bear).

Although I (a minority of one) may argue that there's more charm in the personality of goats than of most cows, long days on horseback allow for a deep consideration of ideas that with time and reflection evolve passing fancy into a more profound bit of philosophy. An enduring quality of humility and forced adaptation marks the cowhand's life. That is not easily done, for whatever reason, in the life of the assembly-line worker, the sales rep, or even the baker or traffic cop. In the matter of poetry, long sojourns in nature translate to prolonged opportunities to come up with an idea or a verse or a rhyming scheme and then revise it a whole lot of times until the phrasing, cadence, and humor or pathos is tuned to a fare-the-well.

A bit of modern-age arrogance is a claim that the really talented person can get by without practice. The cowboy poet understands that is hooey; there's a place for "spontaneity," and that's at the privacy of your desktop or testing a few stanzas in front of a mirror. Experience does indeed count for something. The excellent poem is produced by wear and tear, like a comfortable pair of Wranglers or chink chaps or the smooth-turning copper cricket of a spade bit that has had all its edges worn smooth by prolonged slobber and good use. What's necessary is to make a performance *seem* effortless, and as anyone knows, pulling off such an act takes an almost eternal stretch of practice. Such are the contradictions of art.

Who are "We"?

But let's mull over for a few moments the question of why cowboy poetry has such resonance for some people. I attended the first Cowboy Poetry Gathering in a kind of accident; I was going through the area and there it was, and in those days when folklorist Hal Cannon was getting things up and running, the concept of "tickets" and "sold-out" had to have seemed hallucinatory. How Elko (and suburbs) has changed since! And isn't that itself a part of the larger story? The audience—and the proliferation of other poetry-reciting venues—is itself simply startling.

Owen Ulph (1916–2004) was the blithe spirit featured in several ranch-based books he wrote that are recognized classics. The owner of the Beartrap Ranch in Lamoille, Nevada, he was also a longtime professor of humanities at Reed College, and in more than a few respects an archetype of the cowhand-professor-rancher. (Photograph by Paul F. Starrs, 1999)

We—and that is a sturdily global "we," since Germans, Spaniards, Venezuelans, Magyars (Hungarians), Mongolians, and Mexicans favor cowboy life hardly any less than "Americans"—face a changing world. What's changing most of all is our contact with the world around us: our cars are hermetically sealed demiworlds where the outside barely intrudes; many people travel interstate limited-access freeways that hammer us with relentless sameness. Riding miles in the back of a pickup truck exposed to the elements is not only rare, in many states it's actively illegal, and only a fraction of 1 percent of Americans have ridden a horse for more than three or four hours a month.

Like it or not, upward of 95 percent of Nevadans, Californians, and Arizonians, and nearly that many Utahns are "city people," according to official definitions of where they live and how many other people they live with. We (that means the city "we,"not the rural "we") have lost a connection to an outside world of nature, a world once feared and once loathed and which was something our farming forefathers and mothers (if they were that and not city people already) would have worked very hard to get away from. Few indeed are the people who have embraced the opposite, though there are some, like the famed Nevada potter Dennis Parks, who, in his lovely book *Living in the Country Growing Weird,* describes a "deep rural adventure" moving with his family from besieged southern California to Tuscarora, in its remote corner of the state. Recollect that at the end of the nineteenth century, the bright lights of the city were taken in the United States to be liberating and exciting; how far we have come, now to wish for a return to the open spaces and challenges that just a few generations ago people were fleeing with élan.

The biggest twinge that afflicts many of us who have worked on both sides of the fence, which is to say, have taught western history, geography, and literature but have also worked for years as ranch hands, is a prevailing wariness: now that "cowboy poet" has become a title that people actually put on business cards and which can provide some significant part of a living wage, there are questions about "authenticity." Cowboy poetry was different in the days when it was a sidelight as opposed to a profession, and therefore the "we" in this is complicated. On the one hand, "we" who admire cowboy poetry are looking in from the outside, but so are some of the poets, despite very earnest attempts (which I know we all applaud) of Cowboy Poetry Gathering organizers to keep things authentic, if that's the right term. On the other hand, the spirit of the writing tries to keep its rural yearning the same, and the poetry is perhaps better constructed and more grammatical and no less fervent if it comes from professional cowboy poets!

A good friend of mine is perhaps the hardest-core field-geography guy I've ever met: the sort of person who spoke of his two terms in Vietnam on long-range reconnaissance patrol as a "cakewalk," and who became an exotic wildlife buyer in Alaska, purchasing smuggled-in rhino horns and civet cat meat and cobra blood and, well, you name it, all the time working undercover for the U.S. Fish and Wildlife Service as one of their ace agents. He's one of the greatest experts on wildlife trafficking in the Western Hemisphere. But what does he really want to do? Well, and I'm not making this up: my friend goes by "The Alaskan Buckaroo," and he's one of the resident experts in Anchorage, Kotzebue, Fairbanks, and parts between in the matter of cowboy poetry.

An upswing of interest in cowboy poetry and the world that it presents is a function of nostalgia, but to be honest and direct about it, nostalgia requires a degree of familiarity with a life that most folks don't have anymore. Teleporting to the past isn't possible, and probably wouldn't be half as much fun as might be imagined if it were. We have, in fact, created false memories, false histories, of ranch and cowboy life that emphasize a few picturesque and dramatic features of cowboy existence (roping, broncs, wrecks, scraps, heading into town, roundups, bars, hats, and horses), but which don't bear too much resemblance to the rote of daily cowboy life. It's hard to find the perfect paean to riding drag, building fence, packing posts, treating scours, pulling nightshade, or trotting for five hours on a ground-eating circle to see how a bunch of cattle are spread out. But some cowboy poets have done just that, and when they succeed it's something to be respected and wonderfully authentic.

"Write Poetry"

Cowboy poetry is performed. It has been argued that this is a twenty-first century thing, that cowboy poetry has a fan base because it's vibrant and done onstage, and that it has eclipsed traditional poetry. That's not true, first of all, because many people do read poetry, whether it's Billy Collins or Gary Snyder or Robert Haas or Louise Glück or W.S. Merwin. And even more write poetry—I have two daughters who do, and they're each better at it by far than me. And that doesn't begin to deal with the slam poetry contests and rap songs or the folkloric wordsmiths like Arlo Guthrie or Bob Dylan or Ani DiFranco, who are performing poets in their own right, but setting words to music. An added energy can be gleaned from conventional (noncowboy) poets when they do readings or performances, but an almost essential trait of the cowboy poet or prose-teller is the need to perform—whether around a campfire, in a bunkhouse, at a roundup, on television, or in the biggest darn meeting hall in the land. Poetry that needs to be performed is a novelty—and a good one.

... About Cowboys

A survey perhaps a dozen years ago found that "cowboy" was the number one career choice of Americans who were left free to elect what they might do with their lives. That came just a year or two after the movie *City Slickers* was released, and certainly is tied to a national paroxysm of enthusiasm for ranch life that came with Billy Crystal's hijinks. There was more to it than that: the cowhand was the contrary figure to the dot-com upwardly mobile, ninety-hour workweek cybernerd. But the cowhand is taken to be beyond that, though I expect that as a career choice, "cowhand" has probably dropped like a rock as experience in the wild is less and less common for Americans, who are struggling to get back to the wild, and tend to do so in odd, if charming, outings such as the Burning Man Festival, which draw the uninitiated out of doors.

In truth, there doesn't need to be a separation of people and the land. It's also wrong to think that ranchers are living in some perfect relationship with nature. They still have to use the Internet, do e-mail, use Excel for their books, pay worker's comp fees, and think about retirement benefits, if not for themselves, then for their employees. Ranch kids want to be able to go to school and talk about *Malcolm in the Middle* or *Desperate Housewives* or even do homework online, so the satellite dishes go up at added expense with bolstered connections to the world. Modernity passes few by. While a brash young cowhand opting for voluntary simplicity may aspire to nothing more glamorous than a custom-built saddle, which can easily cost a half-year's wages, impossible expenses are just a foretaste of what has to be dealt with later in life. While the Ang Lee–directed film *Brokeback Mountain* was controversial for its theme of male love, the movie struck a number of friends of mine in altogether another way: it gave a remarkably accurate portrayal of the often desperate poverty, mere scraping by, of ranch families in the late 1950s and '60s. Everyone has moments of dreaming about being freed from responsibility, of escaping the limits of family and schooling and relatives. That dream does not make it so.

The wonder of the cowboy poet is that those features come back to life. That anyone could ever think the cowboy was a creature of the wild, a citizen of nature, a "free spirit," hasn't really listened to the poems that are recited at Cowboy Poetry Gatherings. They're about community, about responsibility, about doing right by friends, and neighbors, helping the land, and doing good. That is, though some might try to deny it, a thick slice of life in the new millennium. We could do well to listen to the lessons of cowboy poetry; a recent essay by Theodore Roosevelt IV (from the presidential line) was titled "Why Is Rural America at Loggerheads with the Environmental Movement?" and it was—very precisely—about how we do need to listen to what is being said in the country's rural places.

There is a remarkable kind of juxtaposition in this country now. I'm not sure how I feel about it, since the politics that go along with it are kind of weird. Rural is cool, and lots of people don't want to stay in the city if it's possible to get away, at least for a while. Second homes are proliferating like tumbleweeds, and the fundamental problem is that I want one, too. This is putting us in a difficult bind that is an expression of what Arthur Ocken Lovejoy once called "primitivism": the conviction that a previous time and way of life were innately better than the life someone is leading now. We can think about turning our backs on the present, but the past doesn't pay those mortgages or pay for our kids' camps, and the alternatives are not great. I think of Jerry Jeff Walker's great song, "The Night Rider's Lament," with its final chorus mocking the outsiders: "Why do they ride for the money / why do they rope for short pay / they ain't gettin' nowhere / and they're losing their share / they all must be crazy out there."

The bottom line is a simple one, although in the light and dark there's a lot of inbetween. Cowboy poetry is about relationships: with one another, with nature, with domesticated animals that depend on us for their feed and to help them and do right by them. It's also about camaraderie and affection, about hard work and going to sleep tired. And reading it—and more important, seeing it performed—is about our modern life and what we can make of it. Cowboy poets try to clear that way for us, and the road is a fine one.

And let there be, in the end, some symmetry: the same Kinky Friedman that we started with had his own conclusion about classic roles in western life: "The notion of the cowboy," he wrote in the *New York Times* op-ed piece, "has always been one of America's most precious gifts to the children of the world." I would be loath to disagree, but I would add: Let us be aware that while poetry captures a spirit, it's in ranching where the real work abides. We must do what we can to keep that alive. The facts of ranching lie in the work itself, and that goes beyond good and skillful words.

James Galvin is the author of six books of poems and two prose books and teaches at the Iowa Writers' Workshop. He still holds on to some land and some horses near the Colorado-Wyoming line, what some may consider the "edge of America," where it helps to know what one is looking for.

A Poem from the Edge of America

James Galvin

There are ways of finding things, like stumbling on them.
Or knowing what you're looking for.
A miss is as good as a mile.
There are ways to put the mind at ease, like dying,
But first you have to find a place to lie down.

Once, in another life, I was a boy in Wyoming.
I called freedom home.
I had walked a long time into a high valley.
A river cut through it. It was late,
And I was looking for a place to lie down,

Which didn't keep me from stumbling
On something, believe me, I never wanted to find.
It was only the skeleton of someone's horse,
Saddled and bridled and tied to a tree.
When I woke in the morning it was next to me.

The rider must have wandered off, got turned around
And lost. It must have been winter.
The horse starved by the tree.

When we say, *what a shame*, whose shame do we mean?
In earnest of stability water often rages,

But rivers find their banks again, in earnest of the sea.
This ocean I live on can't hold still.
I want to go home to Wyoming and lie down
Like the river I remember with a valley to flow in,
The ocean half a continent away.

The horse I spoke of isn't a reason,
Although it might be why.

Page Lambert has been writing passionately about the natural world for fifteen years, finding her own spirituality deeply connected to the land. In "Birth, Death, and Renewal," she gives us an intimate look at the strands of life that form the web that connects us all, comparing her own experience in the West with that of peoples of continents far away.

Birth, Death, and Renewal:

Living Heart to Heart with the Land

Page Lambert

Legend has it that when the ancient Greek hero Hercules engaged in mortal combat with Antaeus, the son of Neptune and Terra—Ocean and Earth—he almost lost the battle. Every time the body of Antaeus came in contact with Mother Earth, his strength was mysteriously renewed. Mighty Hercules slew Antaeus only because he managed to wrestle the giant's body from the land, lifting him away from his source of strength, his very source of life.

When I moved from our small ranch in the Bearlodge Mountains of Wyoming, I felt as if I, too, had been torn from the earth. Severed by a Herculean destiny from all that sustained me: from our beloved border collie, from the horses and white-tailed deer, from the raucous blue jays and red-tailed hawks. For me, as for Antaeus, the loss of emotional, spiritual, and physical strength was sudden and dramatic. But for how many has the loss been so gradual, so many generations past, that it has not been noticed at all? Does this lack of awareness make the loss any less significant? How has our concept of home changed because of this?

Living heart to heart with the land—any land, anywhere—creates a profound kinship. The old twin ponderosas on the high ridge become the grandparent trees. The three-legged porcupine who gnaws the sparse grass in the hay field each evening at dusk is wished safe passage. If the

149

lone pronghorn who stands sentinel near the lone cedar tree is not there, he is missed. The draw that holds the bones of the mare you rode as a teenager and the bones of the newborn calf your daughter could not save becomes a family cemetery. The rocky hill where the lightning struck becomes a prayer knoll.

The land has become the village where you live your life and as such, must be cared for, kept intact for the next generation, and the generation after that. When the creek that once ran no longer runs, when the willows die back from the edges of the pond, when the painted turtles can find no soft mud, one must think of the grandchildren, born and unborn. The land is resilient, but she must rest if she is to remain fertile.

When I was four, I did not understand the word "homeland" but knew the pine scent of Colorado's mountain wind, the feel of her rocky soil beneath my feet, the rocking rhythm of our paint horse beneath my thighs. By the time I was ten, after our fourth move, I understood the angst of uprooting, yet I rediscovered a sense of belonging when I walked the sandbars of the Platte or sat sequestered in the cottonwoods that surrounded the frog pond.

At thirteen, I traveled with my family across the Atlantic Ocean. During the next twelve months, we set foot on the foreign soil of twenty-six countries. I missed home but fell in love with the world. What was the world, after all, but one homeland after another? Ireland, Luxembourg, Russia; Egypt, Lebanon, Turkey; China, Thailand, Japan. Each a land intimately and deeply loved. I marveled at the highland cattle of Scotland, the free-roaming, sacred white cows of India, and the press of humanity in Hong Kong.

A year later, our travels came to an end as we sailed across the Pacific and under San Francisco's Golden Gate Bridge, stepping again onto American soil. En route back to Colorado, the same river on whose shores I had daydreamed swelled with floodwaters, left her banks, roared through the cottonwoods, across farmland and pastures, and claimed our home.

We moved yet again, this time to a small suburban home on the edge of a plot of undeveloped, still-wild land where I hunted for pet garter snakes. For the next year and a half, I saved my chore money until I'd accumulated enough to buy a young, strawberry roan mare. I bridled Romie, slid on, and rode back to the shores of the river that I loved. The frog pond was gone. Driftwood lay scattered on the sandbars. The upper portion of a horse's skull lay embedded in an undercut bank. When the flood of 1965 hit, some of the horses at the racetrack and the fairgrounds were unable to escape. Within a few yards of the horse's skull was a cow's skull. Not all of the ranch animals had made it to safety either.

Eleven years later I married a cowboy, the fifth generation of his family to be reared on a ranch in Douglas County. By 1976, Douglas County was becoming our country's fastest-growing county, on its way to having our country's highest median household income. Ranchland was being developed at such a rapid rate that land taxes skyrocketed and houses took over thousands of acres of grazing land. Well companies drilled deep into the aquifers to meet the growing demand for water, setting the stage for a growing *lack* of water. Land-use plans came too late. We helped round up cows and calves at one of the last brandings held on native grasslands soon to be ripped up to make room for more than 100,000 homes. The subdivision's current covenants, I understand, don't permit basketball hoops in the driveways or birdhouses in the backyards.

When unresolved heir conflicts forced the sale of the Lambert family ranch, my husband and I sold our own small place and, with a three-year-old son and a six-week-old baby girl, migrated to the Black Hills. After two years, we bought a small orphaned ranch in the Bearlodge Mountains of Wyoming, transplanting six generations of ranching roots. The purchase included a ten-year lease on a state-owned school section that was fenced within the deeded land. We had suddenly become the stewards of land held in trust for the citizens of Wyoming.

We drilled a shallow well, built a modest log home, hung our saddles and tack in the 100-year-old barn, and I began writing stories of our new life. Mark returned to Colorado to gather the last of our belongings, loading our horses into a wood-slatted stock trailer. Romie, by now, was an old mare of twenty-four and I was a new ranch wife of thirty-five. Mark and I bought a few head of 4-H sheep for the children, began saving for our first few cows, and within a few years had a small herd.

These were the most embattled years of the war between radical environmentalists and ranchers. The citizens of my beloved homeland were still enamored with cowboys but fell victim to political hype and oftentimes unfounded accusations. Our largely urban nation became unenamored with cattle. "Cow Free by '93!" shouted the radical environmentalists. "Ignorant yuppies," grumbled the old-timers, "don't understand a damn thing about the land."

I agonized over claims by environmentalists that blamed small ranchers for the profit-based decisions of corporate agriculture, and watched my husband, who worked as a range technician with the Forest Service, spend more solitary days down at the barn. On my days off from my bank-teller job, I pored over conservation science and read discouraging reports on the economy. I wrote panicked letters to the State Land Board when a local teacher tried to gain control of the school section. Our small ranch could not support itself without those grazing rights. Mark and I traveled to Cheyenne armed with all the facts and pleas we could muster, plus letters of reference applauding our care of the land. But the pressure on the governor to increase revenue from school sections was mounting. A few sections in Jackson Hole had even been sold, bringing top subdivision dollar. And the legislature was feeling increasing pressure to remove preferential rights awarding state leases instead to the highest bidder (the ones with the most money, not the most stewardship).

Miraculously, we returned with our lease intact for a few more years. We breathed a temporary sigh of relief and continued to build our small herd. I hiked the high prairie, found solace in the deep draws, and explored the shadowed deer paths. I looked to the animals for answers, and wrote. When the crippled calf Broken Tail was born, I could not help but wonder if she embodied the future of traditional rural life in America—a way of life endangered by hungry developers and well-intentioned but poorly informed environmentalists.

Yet our own offspring thrived. Matt and Sarah ran up and down the deer paths and explored the high ridges. They gathered the sheep off the hay field before nightfall, howled at the coyotes before bedtime, fed the weaned calves before daybreak. They saved their 4-H money toward college tuition and weanling colts. They helped old cows die and young ones be born. They learned the lay of the land intimately: her meadows and oak forests, her rolling hills and steeply rooted ponderosas. By the time they were teens, they understood homeland and the concept of a shared landscape clear to the quick of their nails and the marrow of their bones. They knew her seasons of long bitter cold and brief summer green. They began to understand how one devastating season of drought can follow another. They came to know themselves, learning about the world within and the world without.

One day, Matt came home to join me beneath the oaks where a ewe had just given birth to triplets. Standing side by side, we watched the ewe eat her afterbirth. The silence hollowed out a safe place for conversation between mother and son, and soon Matt was asking me questions of birth and maturation. "What happens to people's afterbirth when they're born?" he asked. "What happened to yours?"

The question gave me pause. "The hospital threw it away, I suppose, but it didn't used to be that way." And then I told him a story from Marilou Awiakta's *Selu: Seeking the Corn-Mother's Wisdom.* Her Appalachian grandfather, upon the birth of each of his five children, lovingly buried his wife's afterbirth in the Tennessee woods in honor of the fact that "we come from Mother Earth, and when we are old we return to Mother Earth."

The ewe and lambs were tied by birthing blood to that piece of earth beneath the oak trees. Matt and I, standing there that day, became linked through story to each other, to another family, and to a new landscape. We no longer simply stood beneath the bur oaks of the Wyoming Black Hills but had set foot also in Marilou's Tennessee mountains.

Yet despite all the faraway countries I had visited when I was Matt's age, despite all the eye-opening experiences in cultures so different from my own, despite learning that human beings everywhere want only to love and be loved, to care for their young and pray for their dead, as a teenager I did not yet carry within me the deep lessons of the land already ingrained in my son and daughter.

Six years into the drought—with the stock ponds dry, the hay fields barren, and our marriage suffering along with the ranch from political and financial pressures—we received notice from the State Land Board that once again a competitive bid on our lease had been turned in. This time the amount offered was extraordinary, nearly twenty times what a small cow/calf operation like ours could afford. But in order to keep the lease, we had to agree to match the high bid. The fact that we had voluntarily cut the number of cows grazing the school section because the drought-stricken land needed rest did not matter to the state. With even less income from the ranch than before, we dug deep into our savings and signed the lease.

In 2001, our county received the official "disaster" designation. And again in 2002. That fall, we made the anguished decision to sell our small family of cows. The grief overwhelmed us. But we were not the only ones grieving. All over the West, ranching families struggled with the economics of foreign politics and domestic policy, and with the ecology of not enough rain and too many animals.

When the Wyoming Stock Growers' Association issued a plea to their members to send in stories of how the drought was affecting them, I sent a copy of a letter I had written to a close friend back in Colorado, who also lived on a ranch. Early on the morning of October 4, I received a call from Jim Magagna, president of WSGA. He was in Washington, D.C., and about to step onto the floor of the U.S. Senate in an attempt to obtain emergency relief funding. He wanted to put a human face on the suffering and asked my permission to read the following excerpt from the letter:

> *How does one tell the animals that there is no grass, no water, no hay, no way to shelter them from the coming storm of disillusion? How does one hang on to hope when all one sees is the frightened look in their eyes as they are unloaded in a strange place, with strange people and strange sounds? How can I say how sorry I am that there is no water in the creek where water has always flowed? No water in the tank that has always been filled? How do you explain to these gentle creatures that you can no longer care for them?*
>
> *Look them in their brown eyes. Tell me what to say. Tell me what to say to the husband who cannot make the grass grow, but thinks he should be able to bring on the rain. Tell me what to say to my son and daughter. Tell me, even, what to say to the faithful border collie who stares at the empty fields and wonders why there*

is no beloved work to be done. Tell me why so few Americans un-
derstand this kind of heartache?

Eventually, a few western states did get federal aid, but most of it
went to states with large crop losses and large operations, not small family ranches. Had Mark and I applied for aid, we would have received
only a few tons of hay—barely enough to feed a small herd for a few
short weeks. Not enough to save them, and not nearly enough to nur-
ture a marriage back into bloom, despite topsoil rich with respect and
admiration.

Why do I share this personal story of triumph and loss? What can be
learned from it? How many other ranch families have suffered divorce?
How many have had to leave the land because of economic and political pressures? And, equally important, why should the American public care? Hasn't 98 percent of our culture already *left the land*?
 Chickasaw poet and novelist Linda Hogan writes in her essay "The
Great Without," "Nature is now too often defined by people who are
fragmented from the land. Such a world is seldom one that carries and
creates the human spirit." She goes on to say, "Soul Loss is what happens as the world around us disappears."
 When I made the hard decision to leave the ranch, I experienced
this loss of soul. I am still regathering the parts of myself left behind. Do
I believe my son and daughter have experienced less "soul loss" because
they grew up on a ranch and are still tied to it? Yes. Do they have a
stronger sense of kinship with the land? Yes. Do they feel more responsibility toward the land? Yes. Is their spirituality tied to the land? Yes.
Does the ingrained belief that they are *a part of nature* inform their de-
cision-making? Yes. Does our nation *need* these informed opinions?
Absolutely. Do I believe that wildlife and land conservation can exist in
harmony with ranching? Without a doubt.
 Not only does our nation need the opinions of those raised on the
land, but the world urgently needs the wisdom of *all* cultures who live
their lives in sync with the seasons of nature, native people who plant

their crops and birth their animals by the light of the moon. Ranching families in the United States are not the only casualties of the misdirected but commendable desire to conserve our world's wild and open spaces.

I recently read two statistics in the article "Conservation Refugees," by award-winning journalist Mark Dowie, that astounded me, pulling me from the intimacy of my own losses and out into the larger landscape of the world. Over 12 percent of the earth's land, an area equal to the entire landmass of Africa (11.75 million square miles), is now under conservation protection. And second, more than ten million indigenous people have been displaced in order to "conserve" these lands (some estimates go as high as fourteen million). These people were not able to transplant their roots to new land, as was my husband. The newest threat to the world's native people, according to Dowie, has become the world's environmental organizations, backed by thousands of corporate sponsors. Increasingly, the native people are becoming more vocal. Dowie quotes Maasai leader Martin Saning'o when he spoke at the November 2004 World Conservation Congress in Thailand. "We are enemies of conservation," declared Saning'o. Dowie goes on to write:

> The nomadic Maasai, who have over the past thirty years lost most of their grazing range to conservation projects throughout eastern Africa, hadn't always felt that way. In fact, Saning'o reminded his audience, "… we were the original conservationists." The room was hushed as he quietly explained how pastoral and nomadic cattlemen have traditionally protected their range: "Our ways of farming pollinated diverse seed species and maintained corridors between ecosystems." Then he tried to fathom the strange version of land conservation that has impoverished his people…. "We don't want to be like you," Saning'o told a room of shocked white faces. "We want you to be like us. We are here to change your minds. You cannot accomplish conservation without us."

These words could as easily have been spoken by any of our Wyoming ranching neighbors desperately trying to defend their way of life. Saning'o, like Linda Hogan, speaks of soul loss. The fact that conservationists and ranchers *both* love the land makes this loss even more heartbreaking and unnecessary.

Dowie, in concluding, makes the point that "many conservationists are beginning to realize that most of the areas they have sought to protect are rich in biodiversity precisely because the people who were living there had come to understand the value and mechanisms of biological diversity."

Though the people he speaks of live halfway across the world, he, too, could just as easily have been speaking about the Wyoming ranchers who are our neighbors. Like Saning'o, the nomadic Maasai who believes that the world's environmental organizations cannot "accomplish conversation" without the knowledge of his people, I believe that we cannot accomplish conservation here in America without honoring the generational knowledge of the ranchers and our own indigenous American Indians, who have lived and *still* live on the land. Conservation *must* be our shared goal.

Should we find fault with those wanting to leave urban areas for their own piece of rural heaven? Of course not. We should encourage them—like Antaeus, they are recognizing that nature is the source of their physical and spiritual strength. And we should applaud those cities that incorporate nature into their urban designs. Should we ask of developers the same level of stewardship and long-term vision that is required of ranchers who want to leave the land intact for their grandchildren? Absolutely. Should urbanites who move to a five-acre horse property in the country be willing to attend a few noxious weed meetings? Of course. Should we discourage developers from selling ranchettes and instead encourage clustered housing (which requires fewer roads) and communal stables (which leave open space for everyone)? Without a doubt. Imagine a thousand acres responsibly developed: two hundred acres of homes gathered into a village, eight hundred acres of adjacent grassland and forest for horses, cattle, elk, coyote, fox, and deer.

The desire to return to the land is not a bad thing. It is a good thing.

The desire to consume inordinate amounts of our nation's natural resources (water, timber, fuel) in order to build a "country estate" (oftentimes a second or third home) is not a good thing. Homes on the original Lambert Ranch now sell for nearly $2 million, and 10,000 square feet for a family of four is not unusual. Why, I wonder, when the majority of the world's women still haul their own water, does a single American family require six sinks?

What *does* the future hold for the land? Will state legislators begin passing laws that protect the open spaces on their public *and* private lands so that state land boards aren't under pressure to sell public lands for short-term gain? Who *will* inherit the ranches? According to Dave Theobold, a geographer at Colorado State University, "in this generation, the West will be shaped by the answer to that question, and by the public's willingness to act, in a sense, as investors in private lands. That means paying private landowners to protect public values—like open space, wildlife habitat, and access to public lands—and to not develop their land." Will the American public be willing to buy locally raised food, cutting down on our consumption of foreign fuel?

For most young people reared on small family ranches, the choice to return to the ranch after college to raise their own families is not an option. There simply aren't enough resources. What will this next generation do? How will they put to use the lessons they have learned from the land? Can we reenvision our economy to enable the changes needed to keep land and families intact—to allow our youth to return to the land?

Author Richard Louv, prior to the publication of *Last Child in the Woods: Saving Our Children from Nature-Deficit Disorder*, did a Google search on the term "nature-deficit disorder" and found zero references. A few weeks after the book's publication, more than 15,000 references showed up. Since then, there have been as many as 180,000. Clearly the public is searching for solutions. Yet here's the irony: while our nation searches for ways to return its children to the woods, we're making it impossible for those already on the land to stay. Our federal government and state legislatures pass laws that protect large corporate entities and special-interest groups at the expense of the small family rancher and farmer. The state's decision to sell our school section lease to the highest bidder cast a black shadow on my children's ability to believe in the family ranch as a viable way to earn a living. Even Wyoming, a state proudly boasting of its ranching heritage, doesn't seem, in their eyes, to support those making a living on the land.

Facing this reality, Matt decided to pursue a biology degree and an aviation career. To help cover college expenses, he teaches flying part-time and works on a farm outside of Billings, helping raise crops and cattle. Sarah is pursuing degrees in animal science and range ecology. She works summers for the Forest Service, putting in long hours out in the woods, or when random fires flare, toiling alongside firefighters to save the grasslands and pine trees of the Black Hills. Ask them where "home" is, and they will tell you it is where the colts they raised still run, where the dog that raised them is buried, where their memories live, where their father mends fence and sprays weeds and awaits the return of heavy snows and wet springs. And the snows will return. Despite drought and divorce and displacement, we live in an abundant world.

And me? I returned to the Colorado mountains of my childhood to care for my terminally ill mother. Last fall, as I hiked the trail that passes by the old picnic grounds where I played as a girl, I came upon eleven bull elk grazing on the slope of a hill less than thirty yards away. A few evenings later, a cougar crossed the road and disappeared into the brush.

Four weeks ago, Matt and Sarah helped me scatter my mother's ashes out in the forest. We hiked less than a mile from the house, then stood on a family of granite boulders and broadcast her ashes into the wind, where they left a fine layer of dust on the pine needles and kin-nikinnick. I gently took two fragile, empty halves of a tiny blue robin's egg, which I had brought with me from the ranch in Wyoming, and placed them on an outcropping of rock, then sprinkled the last bit of ash beside the empty shell. Birth and death. And in between, living heart to heart with the land, wherever home might be.

Mark Spragg is the author of three books. Where Rivers Change
Direction, *a memoir, won the Mountains and Plains Booksellers Award,*
The Fruit of Stone, *and most recently,* An Unfinished Life *are both
novels. All were top-ten BookSense selections and have been translated
into fifteen languages.*

Wintering

Mark Spragg

I drive home to northwestern Wyoming with the car windows down.
I do not play the radio, or sing, or whistle. I listen to the wind. I do
not speak to the wind. I have had to explain myself too variously, too
specifically. A gas station attendant in Thermopolis asks if I want my oil
checked, and I turn and walk away from the man. There have been too
many questions. The wind asks for nothing. I have worn away from
conversation, television, music, even laughter. I want to be in the mountains, hold myself against them noiselessly, and mend—inside and out.

I have just graduated from college and my legs ache as though I've
been required to stand for four years, thigh deep, in a pool of December
river water. I've been told that brain matter does not experience pain. I
have no way to assess the damage there. My legs throb so badly I wish
at times to lose them. I grimace when I take each step.

I rub my thighs as I drive. I reach lower and squeeze my calves. The
muscles feel swollen, thick, cramped. I suck at the wind. I wonder if
tennis shoes would have helped. The thought makes me smile. I own a
pair, but wear them exclusively inside a pair of waders when I go fishing. I did not fish in college.

On the flats south of Meeteetse there are antelope. Two, and then
three bunched together. The sage is grown up to the animal's bellies, lush
aqua and muted grays. The antelope look to be boats bobbing in the

chop of a northern sea. Boats without fishermen. Boats of beauty, their bowsprits carved into pronged horn, catching the sun, simply drifting.

Some of my classmates wore hiking boots; the kind of boots that a climber might purchase to tramp through the Himalayas. Most of them had bright red laces. I thought they looked ridiculous. I did not see myself as a climber, or for that matter, a walker. I saw myself as a young man temporarily afoot, without a horse. I wore what I have grown up wearing. I wore cowboy boots. The college was in Laramie, Wyoming, so that was not unusual.

What was unusual, for me, at least, was the concrete: sidewalks, parking lots, roads, paved squares. My legs have lengthened off-road. They are accustomed to soil, wet and dry, the plush of meadow grass and pine needle. Four years of college on concrete has nearly crippled me. Graduate school would have likely landed me in a hospital. I limped away from Laramie swollen from my lower back to Achilles' tendons. I did not attend graduation. I packed the car and did not look back.

I stop the car outside of Cody. The sun is setting behind Rattlesnake Mountain. There is a low scatter of exaggerated cumulus come bronze and mulberry and vermilion. I sit on the road's shoulder and close my eyes and breathe home air. The wind lifts my hair and releases it against my forehead. I feel the mountain's presence as palpably as a change of season.

For the next two months I work odd jobs—shoeing horses, building fence, some carpentry. I live with my parents. I am not good company. By the end of July I find a job managing Elephant Head Lodge. It is not far from where I grew up. I think that if I can make it through to the end of summer, watch the aspens turn and the tourists turn for lower land, I will be all right. Only a month or two of smiling at strangers and I will drain the water out of the cabins, strip the beds of linen, get the horses on winter pasture, and put up a *CLOSED* sign. Two months, I think, and I will be out of the business of interaction, and in the mountains alone.

I dream of January. Of dusk at four in the afternoon. Of the careful, slow pace of below-zero living. I imagine that I will wrap myself in winter: five months of gauzy light; the sweet stultification of the cold. I close my eyes and think of the creeks and river capped with ice. And the snow. Day after day of soft accumulation. Snow that will blunt all noise to whisper. I imagine that walking on snow will mend my legs, massage my psyche open to a pliant balance.

By the first of September the tourists are thinning and it freezes two nights in a row. In the mornings I hear a bull elk bugle. My excitement gathers. I feel the momentum of my prayers for solitude gaining mass. Fred Garlow calls.

He wants to know if I will caretake a ranch. He tells me an eastern couple has bought a place, built a lovely home, and wants a man to watch it.

"Just watch it?" I ask.

"It's the Four Bear. There's some wild horses."

"How many?"

"I haven't gotten close enough to all of them to make a count."

"What do they want done with them?"

"They want them kept wild."

"Isn't that Olive Fell's place?"

"She sold it. She's got life tenancy. So do the horses. She's been up there forty years by herself. With only the horses. I don't guess she wants company any more than you probably do. The place you'll be living is a mile away."

"I have a job," I tell him.

"This one's better. It's year-round and there's no dudes in the summer. Tell me you'll take a look."

"When?"

"Tomorrow."

I meet Fred where a dirt road breaks off from the two-lane that runs from Cody to Yellowstone and get in his pickup. He drives alongside Jim Creek for a quarter of a mile and then switchbacks up a sage and sandstone ridge for another two miles. We rise off the valley floor slowly. I step out to open a wire gate. "I can see into the park. That's thirty miles," I tell him.

As I walk back to the truck I keep my eyes toward Yellowstone, watching a mass of storm roll off the Continental Divide.

"We get up by the house, you can have a look into the Southfork. Over the top of Table Mountain. You won't want for scenery."

The switchbacks become tighter, steeper, the roadway canted. "Think this gets slick?" I ask.

"I think it can be a soupy son of a bitch. There's a lot of bentonite. If I were you I'd stay home when it rains or you'll have gumbo to your hubs. It's the shits getting stuck going downhill."

We top out under a log home built on an exposed knob. Floor-to-ceiling windows flash on three of its sides. Fred squints into the sun, looking up at the house. "That's Olive's place. She got a crew of Finns down from Red Lodge to build it for her in the early '50s."

There are ravens perched on the peak of the roof.

"I've never met her."

"I imagine you've heard that she killed a couple of people. A lady who lived with her one winter, and a hired man."

"I've heard that about everyone who lives alone."

"I was with the bunch that rode this place. We found the body of the woman."

"I heard she died of exposure."

Fred spits out his window. "When you're dead you're dead. If you die outdoors in the winter, I guess you're exposed."

The road curves along the uphill rim of a bowl for another mile, ending against the lower bulwarks of Jim Mountain. "This whole basin'll come up in wild iris in the spring," he says.

I nod. I try to imagine splashes of purple and pale blue pocketing this open, tan landscape. The new home is sprawling and angled and looks to be a broken escarpment fallen from the palisades that rise above it. It is constructed entirely of wood, native stone, and glass. Sage and wild grasses grow to its sides. Fred steps out of the truck. "This is it. Something like eight thousand square foot I'm told."

"They expect me to keep it clean?"

"We can probably get one of those services in town up to clean it. I think the folks that own it will be out for only a month in the summers."

Fred shows me the walk-out basement apartment where I will live. Its south-facing windows let in so much light I pull down the bill of my cap.

"Are they good people?"

"Who?"

"The new owners."

Fred pulls at an ear. "I like 'em better mostly than people I've known all my life. If I didn't I wouldn't have hauled your butt up here."

I trust Fred. I've known him since I was a boy. I've heard him dismissed in town because he's Buffalo Bill Cody's grandson, but consider that just bullshit from gossips with enough spare time to worry over their neighbors' lives. The old cowboys who have worked with him estimate him honest and a first-rate hand.

The upstairs of the house is quiet and crafted with the care of a church. The walls are hung with Russells, Remingtons, Wyeths, Bierstadts.

I stand in front of a huge Wyeth canvas. "This one of the reasons they need someone on this mountain?"

"One of 'em. These things aren't reproductions."

We leave the house and walk west into the shade of the pines and cottonwood that border Jim Creek. There is a set of corrals and several small, dark cabins. "Olive used to live in one of these until she got enough money for the house out on the ridge." He sucks at his teeth and hitches up his pants. "She's kept this ranch together without any help. I guess by selling her drawings in the Yellowstone stores. And some paintings. I don't think I could have done it."

"How high are we?"

"Over seven thousand feet."

We walk back to the truck. I step onto the rise of a boulder and away from Fred. The day is starkly clear. I can see thirty miles to the east, twenty south, and forty west. Without turning I can feel the mountain of Forest Service pine and spruce silently against my back. I imagine I can feel it contracting, cooling, preparing itself for darker, colder months. My legs ache, but the pain is manageable.

"What do you think?"

"You been up here when it blows?"

"You'll get some wind. It'll drift."

I look back toward the basin we've skirted to get to the house. It drains west into Jim Creek. I imagine a westerly wind sweeping it clean, burying the road tucked under the fringe of its eastern skyline.

"If it snows early and stays, I could be in here from November to March. I've just got that little '52 Scout."

"There's a storeroom next to the apartment in the basement. And a chest freezer. I wouldn't let you come up here without putting five or six months of food away. There's a phone if you get hurt or feel chatty, but you might want to say your good-byes."

I step down off the rock. "I'd be alone?" I try not to look too eager.

"I imagine you'll get all the alone you can tolerate."

I turn away from Fred and smile into the searing clearness of the day. "It'll take me a week to find a replacement unless the guy I'm working for has someone in mind."

"Make a list of the groceries you think you'll need. There's a washer and dryer in the house so you'll be able to keep what clothes you've got laundered. I've found it helpful to overestimate on the toilet paper, aspirin, potatoes, and beans. You a hand with the bottle?"

"I guess I wouldn't bring any. I've already been in a poor mood for a couple of years."

We get back in the truck and Fred lights a cigarette and thumbs his hat back away from his face. "I'll call Olive and tell her you're coming up here. I'll let you introduce yourself. She doesn't hold up well under company. You understand?"

"I don't know that I do."

"If you get real lonely she probably won't want to go dancing."

My father tells me flatly that what scraps of sanity I have will be gone by February. "A caretaker?" he asks and shakes his head. His voice rises. "You went to four years of school to be a caretaker?"

"It's not a career choice. It's for the winter."

"It's a job for old drunks or crippled sheepherders."

"What about Christmas?" my mother asks.

I tell them that I think the coming weekend would be a good time. We exchange gifts on Labor Day. They give me a down vest and a pair of lined workgloves. I've framed a drawing of a bear I've made. My mother hangs it in their bedroom.

"I hope it works out for you," my father says. He's settled from the shock of my announcement. We shake hands. My mother gives me a hug. She holds her hand against my cheek and stares into my eyes until

I look away. She kisses me, and hugs me a second time. I tell her I plan to unplug the phone at my new home. She looks down to the kitchen floor.

"I'm not mad," I tell her.

"I wish you were."

"Why?"

"It'd be easier for me to understand."

"I love you," I say, but it doesn't feel like enough.

She nods. Her arms are folded across her chest. I hug her and her arms drop and hang between us.

I drive up the mountain the next morning. The fall is holding, the days warm, the nights chill, the trees are beginning to turn, their leaves ripening to bright yellows, golds, and russets.

I park next to the house and stand out of the Scout and can't believe my luck. I seem above most of the world, above all traffic. The air is as blue as lapis. It will fade to the color of frost in a few short months. I have been offered a home where winter will last. The snow will remain untracked until I walk it. I unpack my clothes, snowshoes, my saddle, books, a typewriter, and a week's worth of groceries.

I lie in the tall grass in front of the house and stare into the still and flawless sky. The earth is warm. I fall asleep and wake in the late afternoon. It does not occur to me that my fantasies of solitude are naive. I do not know, for instance, that fantasy and intent, at a high lonely place, work themselves against the conjurer's mind, their edges knapped sharp and dangerous as obsidian knives. I do not know how permanently I am cutting myself away from the life I have lived, and that at altitude all wounds heal slowly.

I fix a light dinner and go to bed after sunset. I wake in the night and walk out to sit under a waning gibbous moon. The air is cool, dry. There are no insects. Owl-sound loops through the half-light like the warming reed section of an orchestra. The yips and howls of a pair of coyotes bring clear the face of a friend killed in Vietnam. His expression is one of surprised wonder. I lower my head and pray that his soul finds comfort and give thanks that I was not chosen to go to war. My breath sounds loud as the stirring of spruce boughs. There is a faint light in one room of Olive's house and that is all. The insubstantial moon has set.

The stars come brighter. I go back indoors and sleep the rest of the night.

I wake early to an alarming and constant scream. I do not say to myself, "Those are the screams of a woman, or bear, or horse." I say nothing. I can feel my heart quicken and rev. I think for a moment that the earth is quaking open, that I am caught in the jaws of some vast accident. I leap from bed and pull on my pants and boots and run from the house. I run getting into my shirt. The sound seems supernatural, as maddened as the shriek of an angered god. I run toward it because I am convinced I cannot escape.

My mind plays through its brief catalog of casualty. There is the scream of a man pinned under the broken and dropped block of an oil rig—a scream of intense pain and bewilderment—a mountain lion shot through the bowel, a male bald eagle broken and falling in what I imagined at the time was a lost fight for family or sex. I know them now as just one sound: a sound that strips muscle from bone and leaves the listener merely skeletal and vibrating, stark and white as an ivory tuning fork. I run faster. I think of some crazed anchorite laboring against the weight of his solitude. I think that that is what I might find. I think of the scream as simply prayer gone wild, uncaged, sheering a man away from his passions.

The scream pulls me toward its center, contracts and loosens my tendons and ligaments, jerks my knees high into the air, makes me generally elastic with fear, and fast. It is not a matter of wanting to witness the source of the sound. It is a matter of wanting to make it stop.

I do not know it is a horse until I find her. A wild mare, gone delirious with pain.

She has gotten into the cattleguard closest the house—all four legs of her, snapped at the knees. She has thrashed while she's screamed and blood runs from her nose and ears and barely from an eye she has beaten loose from its socket. Her leg bones have torn through her flesh, and work as she struggles like the jagged teeth of some beast feeding on her from the earth, stopped from rising into the air by the iron bars on which she lies. She is losing her fight to live. She breathes hard between screams, her mouth slick with a pink foam, her one good eye rolling white, but glazing rapidly in shock. I think she must have fallen into the thing at a run.

The sight of her turns me faster toward the house. I scrabble in my duffel for a pistol. I find the thing and grip it as though it is the neck of a rabid dog. I grip it hard because I am alone and know what I am about to do. I make my second sprint to the mare. Her cries are deeper. The ground shivers with them. Birds fly toward me in confusion. Prairie dogs scatter in a panic to avoid the slaughter they believe she heralds. I reach her, breathe in once to steady my lungs and heart and fire into her head. I have to shoot her twice, she is that determined in her struggle, and then I walk into the sage and sit. I drop the pistol in my lap and hold my palms against the earth. My shirt is unbuttoned, open, hanging at my wrists.

I begin to howl, perhaps because I am alone. It seems my only adequate response. I do not worry that I will be heard. I howl until I am out of breath. I look around me, blinking. I wonder if any part of me has gone with her, broken through the thin and penetrable surface that holds us away from our individual deaths. I grip my head and chest and thighs with my hands. I make sure that I am still here, complete. I sit quietly for half an hour and then walk to the house. My legs feel weak. My mouth is dry and tastes like steel. I do not feel like eating. I button and tuck in my shirt and pull on a jacket. I find a log chain and bucksaw and back my little four-wheel-drive to her body.

I saw through the ruined bone and tissue and hide at her knees, work her legs from the cattleguard and lay them in the back of the Scout. I pull the length of chain to her. I loop it around her neck, hook it snugly under the ridge of her jawbone. I circle the free end over the trailer hitch and ease out in the lowest gear. The chain bites solidly, her head holds fast, and I drag the amputated corpse a mile from the buildings. I slide her into a deep ravine and toss her legs on top of her. In a week she will bloat and rot and the eagles and ravens and coyotes that catch her scent will come to this steep place for a meal. When I drive back across the cattleguard the tires slip in the apron of blood-soaked ground.

I make a pot of coffee and stall for time. It seems that I've done too much all at once. I check my watch. It isn't yet mid-morning.

I walk toward the creek hoping the sound of water might settle me. I find two weathered sheets of plywood and several stockgates leaned against the back of a small barn. I wire the plywood over the cattleguard and put in a gate. I do not want to trap another horse. This one is enough.

Fred comes at the end of the week with his pickup bed heaped with canned goods, flour, beans, toilet paper, and two hundred pounds of packaged and frozen meat that he's packed with ice and wrapped in a tarp. He pulls a trailer and we unload three riding horses and tie them to the buck-and-rail fence and carry the supplies into the house. "Thought we might get some fence mended before it snows," he says, and later, while we are saddling the horses: "Those cattleguards don't work worth a damn if you board 'em up."

"One of Olive's horses got in it."

"I saw the bloodstains. You get it out yourself?"

"With a chain. I had to kill her."

"Olive see it?"

"She wasn't there."

"That doesn't mean squat. She bought herself a telescope with some of the money she got for this place. I've looked through it. She can see the color of your eyes if she wants to."

"Then I guess she saw it. She hasn't called."

"I don't think she would."

We ride up the creek, climbing out on a ridge that borders the Forest Service. There are three hundred yards of fence down across the crest of the ridge. We reset posts and stretch new wire where it's needed and untangle the spooled mess of wire that the elk and feral horses have walked down. Fred stays the night. We finish the fencing the next day and load the riding horses into the trailer. The days have remained clear. The evening light is soft and grainy with a late hatch of flies and mosquitoes. I ask Fred if he wants dinner.

"I'll eat in town," he says and lights a cigarette. "You all set?" he asks. "Now'd be the time to think of what you might need. There won't be enough work to keep you busy."

"I brought some books."

"I hope you don't mind reading them twice."

"I don't. I think I might try to write."

"You have plenty of paper?"

"I guess I do."

"All right then." He gets in his truck. I don't know that I won't see him again until spring.

Within a month, a rain and two wet snows fade the mare's blood to rust, to ocher, blending it finally into the dark, brown ground. Her scream remains. I am sure I hear pieces of it each day leaked from the side of the mountain. It comes out edged and brittle and wracked as the morning it tore itself loose from her, the mare, a bay, a horse unused to the inventiveness of men. I grow accustomed to it. I think of it as bright red. I hear it as a melody that accompanies my isolation.

In October the phone rings. It is the first time and it is before dawn. I do not know what time it is; I have put my watch away in a drawer. I get up in a panic. It is a woman's voice. "Are you the man in the house?" it asks.

"I suppose I am."

"I'm Olive Fell. There's hunters up by the spring. The one over on the east corner."

"How do you know?" I've found a lamp and turned it on. I'm settling.

"I scoped them. I want them run off. There's no hunting here."

"Right now?"

"I'm an old woman. My legs aren't up to it."

"Yes, ma'am."

The hunters are a local man and his midwestern cousin. They wear camouflage clothing and sit their horses as sluggishly as clods of pond mud. The cousin has sprinkled himself with cologne. I smell him when I step from the Scout. They are belligerent. They are trespassers and indignant that they are caught, but they keep their rifles in their scabbards. They claim they don't know they're on private land. I tell them I've found where they've cut the fence to ride onto the place. I take their names and ask them to leave. I escort them back through the opened fence, then bring up the wires and splice them together and staple them solidly to the posts. They don't say they're sorry. They tell me to go to hell.

When I get to the house the phone rings again. It's Olive. "I should meet you," she says.

"All right."

"After lunch. The day's been ruined anyway. Those men give you any lip?"

"Yes, ma'am."

"Thank you."

"It's okay."

"They kill anything?"

"Not that I saw."

"Thank God for that. And thank you for boarding up that horse trap." She hangs up.

When she opens the door she shakes my hand right away and then turns and walks into the kitchen. I follow her because I don't know where else to go. She is only about five feet high, has short red hair, her body plump and speckled. I guess she is in her seventies. She wears blue tennis shoes, blue denim pants, a short-sleeved blue fishnet undershirt, with a denim tunic pulled over it. Also blue. The pants and tunic are boxy and appear homesewn.

"You want some tea?"

"I'd drink a cup. How did you see them in the dark?" I ask.

"I saw their headlights. Where they parked and unloaded." She puts a kettle to boil and turns and looks me up and down. She folds her arms across her chest. "I never killed anybody," she says.

"I believe you."

"You don't have to."

"I know I don't."

"I don't have guests, but when I do they have a hard time not asking. The only one who doesn't want to know is my brother and he doesn't visit."

"I've been in his flower shop in town."

"Usually takes me about a week to get over seeing somebody. Like when the man hauls hay cubes up for my horses. That's why."

"Why what?"

"Why I don't have anybody in." The kettle whistles and she turns back to it. "Do you know why you've come up here?" she asks.

"I want to be alone. I think I might like to write."

"That'd be a tough way to make a living." She turns back to me with our cups. "You've found a good place to try," she says.

We take our tea into the living room and satisfy ourselves with the view out of the three walls of windows. The sun catches on every surface in the room. It is like standing inside a flame. There is a painting on an easel by the fireplace and tables mounded with prints. She shows me her studio. She takes a shoebox out of a drawer and hands me letters to read. They are from Georgia O'Keeffe and John Steuart Curry.

"I went to the Art Institute in Chicago," she says. "I thought once I might like to marry Mr. Curry."

We talk until it is dark. She turns on a floor lamp in the living room. A bobcat sidles out of the darkness and into the soft, yellow lamplight fallen out the windows on the west side of the house. He rubs his muzzle against the glass. Olive walks to the window and holds out her hand, making a clucking sound with her tongue. The cat grovels against the sill, his eyes half-closed. The light catches in her hair and glows lightly against her freckled arms. "This seem unusual to you?" she asks.

"I've never seen a bobcat act that way before. Did you raise it?"

"No," she says and kneels by the window and presses her hands against the glass. "I've been up here a long time," she says. "I've never hurt anything. It makes a difference."

"I can see that."

"I don't use the horses either. They've gone three generations wild now. Before I could afford a car I'd walk down to the road and get a ride to town."

She pulls herself away from the glass and goes into a storeroom and comes out with a small harness. "I used to wear this," she says. "I'd hook it into a sled and pull my supplies back up the mountain. Soled my shoes with worn tires, wire, and horseshoe nails. Art doesn't pay. You hungry?"

"I guess I am."

"You better get home then. I don't have a good appetite. I mostly just pick. Mostly at night."

I return our tea cups to the kitchen. She is waiting for me by the front door. "You're a good boy, aren't you?" she asks.

"I haven't gotten in much trouble."

"I hope you get your writing done. I hope it makes you happy."

"Thank you."

"I'll call if I spot any more hunters."

I look toward the telescope set up in the living room, facing north-west. Toward the house where I live.

"I was thinking about unplugging the phone," I tell her.

"I'd appreciate it if you'd wait until hunting season's over."

"Thank you for the tea."

"I won't spy on you."

"Ma'am?"

"I just watch the land. It takes some watching. There's a lot of it."

"Yes, ma'am."

She calls once more on Halloween and again the first week in November. Both times she has spotted hunters. Both times I find them and ask them to leave. I tell them they are on private land. I do not tell them that they are interrupting the lives of private people.

The week before Thanksgiving, hunting season ends. I call Olive and tell her I'm going to disconnect the phone cord from its wall jack. She says she understands. I call my mother. She tells me she loves me. She asks if I have plenty of food. I tell her that I do not want to run the risk of communication. She says that I've always been the way I am, that I shouldn't worry about having gotten worse. She says that she will always love me. I hang up and look around the room. I've taken pre-cautions: no radio, no television, no record player, no chance to be tempted back to lower elevations. I vow that I will not lose altitude. My legs tingle. They're stiff. The pain only comes now in the middle of the night.

Every day I walk, and watch, and write on a pad balanced on my knees. I have been hired to guard against frozen water lines, theft; to keep the buildings heated so they will not constrict to thirty-five below. My days and nights are my own. The snows come and drift and close the road and leave me satisfied and unsurprised. The freezer and pantry are full. I come indoors to eat and sleep and bathe, and for the warmth. The days grow shorter. I feel the Earth yaw back away from the sun. The raptors, the coyotes, and the grown-fierce, inbred horses are my com-pany. There are fifty-two of them. I have made a count. They prowl the ridges blown free of snow and congregate in the mornings at an auto-matic feeder housed in a shed below Olive's house for their daily ration of hay cubes. They fight among themselves. The mares. The uncut studs.

I have seen them, twice before the snow, catch a coyote out too far from cover and encircle the quick gray dog. By the sheer force of their numbers they vector every angle of escape. They bare their teeth and lay their ears along their skulls, and strike with their front hooves. They hack the little predator to a mess of trampled bone and pulp. It has made me feel fragile and clumsy to watch them. I feel they blame me for the lost mare. I know they hear her last cries of pain, as I do, daily.

I think it skitters in their minds that I can go the way of a coyote. When I walk out through the sage they become curious, rapacious, swarm around me. Half a dozen times I've bounced an armful of rocks off their sides, and waved my coat above my head, and bluffed the mob of them apart. But when the cold pushes the songbirds south and freezes the fist-sized stones into the mountainside I become more careful where I walk. And when the snow is waist-deep, I walk out on snowshoes in the night with only my mooncast shadow to watch my back. Just the two of us, safe, scribing a blue-shadowed trail across the drifted snow.

There are a few weeks of dislocation and then I line out and write sixteen, sometimes twenty hours at a stretch. I will at times, the best I can gauge, sleep a day and a half and wake in the night and wash and go to work. I make stories. They become true. They become my history. I forget that I have lived on lower ground. I sit and watch the land. Time leaches out of me into the snow and wind, loses its rote and civilized boundaries, becomes a whole and seasonal thing. I grow more comfortably wild. I eat when I am very hungry. I watch the mountains at the horizons for complete revolutions of the planet and observe myself grow increasingly quiet, and more gentle. The untamed lives that surround me move closer. At times, mice, and once a squirrel, sit on my knees, not altogether comfortably, but they come that close—or, I to them. When lonely I write myself a guest. I speak to him and he speaks back to me.

I wonder about madness. About definition. I try to summon my mother's face, my father's, my brother's, the girl I thought I'd loved just a year ago, but they seem too far away to reimagine. As though they have all died and I am left without their photographs. My body remembers emotion and demonstration, but vaguely: anger, affection, even caress seem blunted to my memory. They come as soft tappings on a roof—that disparate. I find that I cannot put heat or perspiration to the few recollections that rise and fade. I wonder if I am to become a man who knows only bobcats and coyotes and hawks. I wonder if the feral horses will someday see me as benign, forget that I am foreign. But I have mostly come past worry. I am content. I am smoothed by the constant work of wind and snow and the dark, cold nights. I feel the season turn in me as tumblers in a lock, leaving me open to the earth. It does not happen all at once.

In the shortest part of winter I become sick. The days are stunned with cold. It is perhaps the end of January. I am not sure how long the sickness lasts or what brings it on. I remember eating from a swollen can, knowing better—I cannot remember why. It was a stew. Bitter and tasting of metals. Mostly I remember the darkness, and the gauzy, brief periods of light that come up full and fall. Three of them I think. I am sick for three days and nights.

I make a nest of towels, a pillow, a quilt on the bathroom floor. I drag them there and vomit into a bucket. I kneel at the sink and sip water from my cupped hands. I struggle on and off the toilet, and lie again on the floor, and shiver and toss against the fever. I am not certain when I sleep. I do not know if I dream. There are times when it feels as though I am falling down a well. There are visions, starkly colored hallucinations. When I think that I might die it does not frighten me. Death does not seem so far from here to there. I reason that madness is simple release, an entrance, an acceptance of my own peculiar struggle. There are hours of sliding dislocation into color and shape. There are chords of sound. I become convinced that I am only an audience and that there are watchers watching me. I am too young to imagine that I have seen the face of any god, and far too young to know that enlightenment shifts and fades. And then I think I sleep.

I feel the mountain throb with the sounds of a dying mare. Her last and desperate scream of pain works in and out of me. Scours me. Loosens childhood memories. Shatters thoughts of the future. Suspends me in each moment, remarkably without sentiment. I wake in the night and am able to walk to my bed.

On the morning of the fourth day I shower and sip some broth. I walk out on the frozen ground, careful that the wind does not blow me east. I feel that insubstantial, and that elemental.

It is two weeks later when I see Olive out and crowded by her horses. They duck and nod and come in to her shyly. She lays her hands on their outstretched necks, their foreheads. She removes her gloves and they nuzzle at her open palms. She is standing on the ridge east of her barn. When the horses see me approach they scatter like quail. She laughs and waves and looks up into the frozen sky. It is cloudless and starkly blue. I come toward her on a drift. Its crust sparkles and crackles, but holds my weight.

"How's the writing?" she asks. Our breaths puff and hang between us in the air, dropping as we speak.

"Better than I thought. It's cold."

"It's winter." She looks southeast and points. "I was raised over there."

"On the Southfork?"

"Between Cody and Meeteetse in a sod dugout. It was a freight stop. My father drove a freight wagon." She looks at me. Her eyebrows and lashes are frosted white. "He was a dirty man. I can't remember whether he was handsome or plain."

"What did your mother do?"

"She took his beatings. Me and my brother too." She looks back toward the place where she was raised. "He beat us all one day until we were unconscious. Then he fell down drunk. I was the first to come awake. We were outside. It was summer. I took up a hatchet and chopped off his hand. I loaded my mother and brother in the buggy and drove them to the county seat. She filed for divorce." She looks back to me. Her face is flushed, but very calm. The thought of a man's hot, red blood seems impossible in the white and frozen air.

"Did it kill him?"

"No." She smiles slightly. "But he never laid that hand on a living thing for the rest of his life. It was a different time. Everybody knew everybody. The sheriff told me I was a brave girl. I started to paint pictures afterward. I felt that free."

"Are you painting now?"

"I don't have the same kind of concentration." An eagle circles over us, flashing in the lightstruck air. We shield our eyes and watch him spiral wider and to the west. "Do you like it here?" she asks.

"More than I thought I would."

"Enough to stay through another year?"

"I don't know."

"If you're here next winter I'd want this feeder checked. It clogs sometimes. I'd pay you for the work."

"I'm already getting paid."

She nods, and cups her hands and blows into them. "I bought a motor home and I've hired a boy to drive it. Next winter I'm going to have him drive me to California. Arizona and New Mexico, too. I want to look at paintings. It's been a long time. I think it will help with my concentration."

"You'd be back?"

"I might not be able to stay away as long as I think. But I wouldn't want to worry about the horses."

"If I'm here I'll do it."

We search the sky for the eagle but he is gone. "I've been worried about the horses," she says. "The winter's hard." She kicks back at the snow with the heel of her boot. "When my father was sober he was a good man," she says, and then, "You've lost some weight."

"I was sick."

"I have been too."

Out that night on snowshoes in the moonlight I lean against a fir to watch my breath freeze and spread. I hear a scratching by my ear and turn slowly and find a great horned owl perched inches from my eyes. He looks down at me and seems to nod, to say in gesture, that for a man I am bending toward a rightness. That is the way it seems to me.

I close my eyes until I can hear him breathe, and open them and the moon has passed behind a cloud, but he's still there. He blinks and ducks his head and blinks again. I back away and move stiffly toward the house to warm my hands and feet. I look for my shadow and find the thing. I laugh. The shadow shakes.

Often, toward the end of winter, I lie in bed and try to reckon whether I have just waked from a dream of words, or whether I have stretched myself down, having tired of a day of making stories in the cold. Once I hear an actual voice. I do not understand the words, but it seems to matter only that I am quiet enough to hear the text.

Close to the vernal equinox the snow is puddling at the height of day and freezing bright at night. I wake and am sure I hear footsteps somewhere over me. Most of the house is built above my bed. Fred has told me that if a thief comes he will be a professional. The paintings are cataloged worldwide. They cannot just be sold and hung. My mind fills with the possibility of a burglary. I look to each side and behind me to see if I am watching, if I've awakened in another dream.

I slip out of bed and creep into the hall. I watch in surprise as my hand thumbs back the hammer of the pistol that it holds. I'd fired it last into a screaming horse. I edge up the stairs and from room to room. I hold the gun in front of me. It catches the light of the waning moon. Each room is sliced into a thousand shadowscapes.

I search the house down to the last room. I stand at its threshold convinced I've notched myself to the nub of death, convinced the assassin-thief I know I've heard waits comfortably for me to walk into his trap and die. I nearly piss myself with fear. I know I am awake and fallen into the world complete. I am afraid of pain. And afraid to die.

I keep my back to the wall and ease into the room and find the man across from me. He's cornered and has raised a gun. His eyes are locked on mine. He waits. He does not fire. In the last fragment of a second before I squeeze the trigger down I realize that the man I face is my twin, and come to sack the place dressed as I am dressed, in his best white, cotton underwear. I slide down the wall, and sit and stare across my knees at the reflection of myself. It is caught in the gathered moonlight of a floor-to-ceiling mirror my employers have flown in from the north of Italy. The house seems suddenly gone empty. Soundless. I cannot even hear my breath. Except for the reflection I have no certain proof that I am there.

I snowshoe out the next afternoon, two thousand feet down to a valley already free of snow. I cut the two-lane at dusk and catch a ride to town with an aged rancher known regionally for his total lack of irony.

"You been up there all winter?" he asks. His voice sounds like the bleating of an animal from some strange zoo.

"I have," I say.

"You should have seen her when she was young."

"Who?"

"Olive. Prettiest girl in the valley. Out of my league." He smiles without taking his eyes from the road, and then looks up at the slice of his face reflected in the rearview mirror. "Still is, I imagine. You get lonely?"

"I'm okay."

"Friend of mine's a plumber," he says and turns and looks at me, checking the road from the corner of his eye. "Olive called him up there a couple of years ago. Her drains were plugged. He got up on her roof to run a snake down the vent. Imagine you know how the wind blows skirting that mountain?"

"Yes, sir. I do."

"He blew off." I look at him and he raises his brows. "Floated to the ground like a goddamn leaf. Back and forth and set down on his feet so soft he didn't have to bend his knees. What do you think about that?"

"I don't know what to think about it."

"You think the wind could swirl around and do that for a man?"

"I guess it did."

He searches my face, and snorts out a laugh. "I guess it did," he says. "You say she's still beautiful?"

"I don't think she cares."

"You don't quit caring."

"I think she has. She's the only one looking at herself."

"Maybe that's the trick," he allows.

We do not speak for the rest of the trip. He drops me at a roadside tavern at the city's farthest outskirt and waves good-bye. "Keep track of how long you're up there on that mountain," he says.

"Yes, sir."

"If you don't, nobody else will."

Inside the bar there is a blaring band and the place is packed. I force my way to the bar and shout my order and think the noise of the people might shatter me. I tuck a six-pack of cans against my ribs and strain back through the crowd and out. A girl my age kneels to the side of the entrance, where the dark begins, and vomits. She drops to her side, asleep, and I cover her with my denim jacket and walk behind the building where there is just a ledge fifty feet across to the lip of a canyon.

I sit on the canyon's edge and stare at my boots dangling hundreds of feet above the Shoshone River. I drink the cans of pale, weak beer. The smell of sulfur rises off the hot springs that edge a bend of the river. The moon is high and on this strand of river, and in a mottled smear of cloud to the south, and brightly to the west on the white and frozen mountain where I have lived alone. I think that once the beginning herds of horses grazed this rim. I think that now there is just this bar with a dangerous parking lot at its back. The night is cool and moist.

I feel the throb of the music drop through into the porous rock on which I sit. It climbs the column of my spine—I think that night—in a single-minded effort to reconstruct me more social. There is, too, making its way into my bones, drunken laughter, voices edged in threat, murmurings of sex. I draw my knees up and hug them and drop my forehead on their caps, my eyes closed, and even though I cannot hold a tune know that what my ears hear in this sad music is the sorry harmony of a dying horse.

I am unsteady when I stand. I start upriver and away from town. My legs feel stronger with each step. I smile into the moon as though it is the open eye of a great horned owl, and I think that if I'd fired into the mirrored image of myself, they would have found the gun and glass, painted black to reflect the light, but they would not have found the place where I had gone. I hold my breath and hear the clean, red scream that holds us all as one.

Linda Hussa ranches with her husband, John, in the borderlands of Nevada's high desert country. Their life and work have taught her about values and traditions, horses and people. "In the Clearing" reminds us that architecture and ranching have an existence that is more than the mere sum of their parts.

In the Clearing

Linda Hussa

The barn lived alone on the land, having passed beyond need. There on *that low knoll, ground sloping away toward water in a shallow creek, trees well back, grass trod out for more than a century, the barn was at rest. I admired its correctness, its manner of belonging, its beauty beyond requirement, and knowing the affairs of barns, I admired the mind that built it, loving it, all of it in the course of daily toil. No fence tied to it, no posts set adjacent to control its line, no corrals holding it down, no other work connected it anymore, freed as a soul is freed from failing breath and stringhalted stride. Yet it throbbed as light poured on it from a broken sky, deepening shadows, brightening cleanly each board from its companions, as a tree stands alone and also belongs to the forest in kinship, in wholeness, in desire. The color of it belonged, blanched pine and iron warming under this light, gray as the grooves of other years in a man's eyes. The sheer plane of the roof, the pitch to shed snow abruptly met the pace and rhythm of the walls' milled boards and bats. It was the flat of endless sky against mountain peaks; it was the horse's flat stride against once being wild; it was the flat gaze of the man set out to change the open country, to mark it by his very being, against his desire to let it remain so. I stopped at the roadside some distance away and stood at the fence, hands placed between barbs, holding wire that still sang of cattle and horses and the dangers of living in a harsh* *land. Even from that distance I knew the boundless wish of the builder to make it sturdy and beautiful and true.*

My grandfather built barns in the wheat country of eastern Oregon, and this one stands out on the windswept plains above the Columbia River in the unattested lines of prairie style. There were others, all elegantly spare. In the Great Basin where I live now, 500 miles away, I regard our own barn, which was built by a stranger's hands. This man, whoever he was, had in his fingertips a single style he repeated the length of the valley. A good barn, and sturdy, it has survived a hundred years in the semiarid borderlands of Oregon, California, and Nevada on the alluvial steppes of the valley's west side. Simply framed and plain, it faces east and corners to storms carried on southwest winds. Still, it was his lifework and serviceable. For some, that is enough.

Built as our house was, perhaps by the same wandering carpenter, twelve-by-twelve-inch timbers rest above the ground on platforms of stone to keep the wood from rotting. Stone bears the weight of the barn, and seems to hold us to the land.

I was born to a horse culture. My parents met at a rodeo dance and married in the spring. They were young and sick to death of families living on cream and egg money and feeling Depression-poor, the stars in their eyes forming constellations of horses. They moved away from the isolated plains of eastern Oregon to train horses for town people.

It was September when they brought me home, still mewling from birth, and the smell of sage came in the open window of the Dodge. Did I know it then? No. But when I met my husband, John Hussa, and he invited me to help gather his cattle from their range in Nevada, sage moistened by the dawn stirred a deep memory. Seventy miles from the nearest town, a hundred years from commuter traffic and a job, I rode into a country I didn't know existed. There I fell in love with cows I would know through the years of their lives, and their heifer calves after them, ranch horses as tough as the lava tablelands, and a man who would teach me his ways and want to learn mine. When we married, I made my life around the thing my parents ran from: a cattle ranch in an isolated valley on the edge of the high desert.

At the height of the Depression, a federal biologist wandered into that same sector of northwestern Nevada where I now live. All but empty of people, the broken land teemed with antelope, mule deer, cattle, horses, and bands of sheep in a flowing pattern like the crosscut of tides. Sage grouse sounding like B-52s flew in by the thousands to water on the meadow, where he unrolled his bed. While the country was mired in financial ruin, he stumbled into a place without breadlines or broken men, as if he'd entered paradise.

He found sheepman Tom Dufurrena, a Basque immigrant who owned the deeded parcels that gave him control over more than a half million acres. At Tom's campfire the biologist told him that if he didn't sell to the federal government, "they'll just take it anyway." Tom didn't understand his rights or the concept of eminent domain, and the lands became public. The U.S. Department of the Interior converted the Dufurrena holdings into the Charles Sheldon Antelope Game Range and Refuge: 34,000 acres in 1931, and an additional 500,000 acres in 1936.

In the year of the final "sale," all livestock were removed from the range. The Dufurrena ledgers record his 1936 livestock census on those acres as follows: 16,000 sheep, 500 cows, and 500 horses. In addition, Frenchy Montero had a cow permit for 500 head and a horse permit, and Harry Wilson had a cow permit for 500 and a horse permit. Now these animals were all gone.

In 1938, the Fish and Wildlife Service (FWS) offered a 1,600-head, cow-grazing permit on that same land to offset operational costs. Three ranchers from Surprise Valley, just over the state line in California, filed an application. John's grandfather, W. H. Hussa, was one of those permittees.

Feral horses had been left to roam when the range was cleared of livestock in 1936, and these lands did not fall under the 1971 Wild

Horse and Burro Act for control and removal of feral horses. The horse population had been increasing by 25 percent a year. Horses were always within sight when we rode that country. They lived free, without fear of predators in the desert. During our fall gathering in the late '70s, I came upon a herd of horses moving across the Rock Springs Table as thick as a band of grazing sheep.

During these years and those that followed, ranchers left their work to attend meetings and toured government study areas hoping to influence grazing policy for the future. I picked up John's father, Walter Hussa, after the final meeting. I had never seen him so depressed. He said it had been an exercise in futility. Many permits were cut by half, some by as much as 80 percent. It was his opinion, and that of many others who endured the ordeal, that the reductions were set prior to the first meeting. Allowing the ranchers to air their objections completed the process.

Every ranch in our valley suffered drastic setbacks during that period. For some it was a total loss. Ag loan companies that had handed out money like feed-store calendars in the early '80s reacted to the panic of the mid-'80s, reversed themselves, and wanted their money back, creating a wake of financial ruin and personal loss across the heartland of the United States.

The interest rate on our operating loan soared to 19.75 percent. The bred-cow market fell as low as it had been since the Depression. The bank officers came to our door. They sat at our table and I put out coffee and cake as if for guests. They examined Walter's neat ledgers using words I didn't understand and eyes I did. When the talk was done, a notice came in our mailbox. The postmistress pointed to the line where I was to sign and turned away. Ask me anything about that day and I can tell it. I remember it the way a bone remembers where it was broken.

Dozens of cattle trucks came up our lane. Their throbbing engines vibrated the windows of the house as they passed by. After we loaded every one of our mother cows, every calf, every bull, we stood in the lane and watched them go.

Cattle trucks loaded with animals forfeited for the inability to pay off debts left this valley like an endless train. It hit the old folks the hardest with a sense of failure they couldn't overcome. That winter we had two or three funerals a week. Our pastor said he didn't think he could stand to sit with another grieving family. But he did.

The next eight years were cut out of our hide. The financial decisions had always been Walter's. Now they were ours. John and I decided that if we were going to suffer the mistakes, we wanted to be making the decisions. We intended to pass the ranch on to our daughter debt-free. The land was all we had left. It had to produce a living for two families. Every expense was shaved. No job was beneath us. We took a contract to calve out 250 first-calf heifers and lost only three calves. We raised drop calves for Oregon dairies. New babies on milk, older ones weaned to grass, the number of bottle-fed Holstein calves we fed twice a day was steady at 300 head. Our ranch was an Escher drawing, all black and white in perpetual motion.

Walter suffered the defeat deeply and spent his days in the shop straightening nails on the anvil. We sold our desert permit on the Sheldon, more than half of our home ranch, and all of our deeded ground in Nevada except for an eighty-acre homestead that was not encumbered. Both of us needed to feel we still had a stake there. As I signed each set of the papers that moved us closer to solvency, I was chilled, remembering Walter's words to me when he handed me a pen and the operating loan application every January. He would push the papers across the table and say, "Here, do you want to sign your life away?"

While we were ears-deep in getting free of debt, all livestock permits on the Sheldon were retired. There was no hope of going back.

To date, not one cow has set foot on the Sheldon in eighteen years. During that time, according to federal studies, coyotes and ravens have all but decimated the sage grouse; mule deer and antelope populations "have been significantly reduced due to several hard winters, and depredation." Three wildfires have raged across the refuge, consuming years-old feed, browse, and groves of mountain mahogany, costing millions of dollars for fire suppression. It is my belief, and others', that harvesting grasses is key to the health of the plant, that antelope and deer benefit from the presence of livestock. Managed correctly, livestock grazing the Sheldon could improve the entire ecosystem.

This fall, the bale wagon backed into the center of our barn and set a block of alfalfa down, which I cut and John baled. We restacked the loads as they came from the field to get more hay inside safe from spoilage. All day long we dragged bales up toward the rafters and stacked them wall to wall. Between loads, hands on the hewn beams, I had the owl's view. How much it felt like resting on a limb with the entire world below. Dufurrena's range, then our range, in Nevada was sixty miles distant beyond the mountains across the playa.

Now our cattle, sheep, and horses run inside on deeded ground under a modest rest-rotation program. We cut the meadows for winter feed and adjust our carrying capacity according to factors we can't control, such as drought, hard winters, and fire. John has taken advantage of fault features in developing stock ponds that extend utilization. As a boy John hiked and played along the creekbed that crosses the ranch. It ran six to ten feet below the level of the fields adjacent. He fished in holes twenty feet deep. Thirty years ago he began building dams to divert spring floodwaters onto the fields. Fine gravel and rich silt filled behind his dam constructions, raising the creekbed and the water table level with the fields. He has improved the native feed greatly. He encouraged his father to stop plowing up meadows, losing the moisture in the soil; to plant oat crops that succeeded only one year in four. The native Great Basin wild ryegrass has returned, giving cover to game birds and winter feed for the cattle. These are small but important improvements to our operation. Livestock and land are unencumbered. We are debt-free.

Standing on hay we harvested and stacked, fingering the cuts on the beam, I know how Grandfather entered my life. His barns, like our ranching operation, are honest, made of everyday labor, made of community, and carried forward by vision. A barn will not stand if it is built on deception and shortcuts. And I know what's troubling me.

I want the government to respect private property rights and work with landowners in honest and fair ways. I want decisions to be made by people who live on the land, close to it—who know it. I want legislation that will support the industry. I want people to recognize that many of the vast lands of this country are best used for the purpose of raising food.

Perhaps most of all, I want ranches to be family-owned, not by wealthy landowners or corporations, neither of which care about the community. Families are needed to care for the land in a way that only those who are invested can care. Invested not by market shares but by history in and love of a place. This kind of commitment comes from a deep understanding and appreciation for livestock and landscape; their payoff is the family unity of purpose, the shared work and responsibility, the ever-present learning, teaching, rewards, and challenges of productivity. The lessons learned are the foundation the youth can build on as entrepreneurs, as legislators, as leaders who protect and renew the environment. It's the way to building and enriching family life in the communities of the West and in our nation.

People still speak about my grandfather in words as well formed, as keen and lasting, as his barns. They say a rancher would explain exactly the kind of barn he needed. Notes of dimensions, dates, and bills of lading filled nickel tablets. No drawings.

Before the rock foundation was laid, they say he cut all the boards out on the ground: beams, supports, rafters, stacks, and piles of hewn timbers and planks. When the materials were ready, he began. A barn lived in his head as music might.

Every job meant a family journey into the forest of old trees and Grandfather's search for one that could be properly milled into a barn. I have a photograph of him standing beside a cut tree with a girth as high as his head. Already middle-aged, and yet he is lean as a boy. He holds the lines of a white horse he calls Robin, and the four eldest of Grandfather's children sit holding one another at the waist. Robin pulled the wagon loaded down with supplies and children to the mountain camp, and snaked logs through the heavy stand to a landing. Grandmother ran the camp, caring for the babies, washing and cooking as she did at home. The older children picked huckleberries or currants or elderberries into lard cans when they weren't occupied with gathering firewood or helping their father. For weeks they lived with the smoke of campfire on their skin and fed on the fruit of mountain thickets and streams. Their voices filled the woods like birdsong lifting above the thump of the axe, whining saw, the whetting stone biting a knife-edge into steel. When Grandfather cried, "Timber!" they were hushed by the melancholy sigh of the giant tree falling, breaking through brush and limbs to the ground, and the cleft of silence afterward as the consort of trees adjusted to new light.

Barns smell of more than fodder, more than animals loafing. They smell of pine and cedar and uprights of juniper cut in winter when sap has pulled to the heart so they will stand and not rot. The smell of meadow hay drying, alfalfa, oat hay, and straw put up for bedding. Tin ticks when it heats and tightens when it cools. Children swing on a rope in and out of light.

Grandfather left no pouch of gold. Riches were the bounty of his imagination and the thrifty use of trees cast by simple tools to create a symbol of community serving both neighbor and beast. He gave me barns, beautiful and lasting. I was one life removed from the guidance of his hands. In time, I learned to hear the music of his barns, and love filled the memory.

Aaron Abeyta has two collections of poetry and is the winner of the Colorado Book Award and the American Book Award. His poem "i remember my father in the future" is inspired by his ranching family in southern Colorado, especially his father's deep spirit and perseverance. Aaron speaks of the need to survive, regardless of obstacles, including economic factors, extreme "environmentalists," and the difficulty of ranch life.

i remember my father in the future

Aaron Abeyta

i am relying on the night for inspiration
mars burns red
on this particular night
our red neighbor is closer
than it's been in 600 centuries
after tonight it will spin away from us
and never again
in our lifetime approach
us with such fire and proximity

my father rises in my mind
the way he always rises
in the darkness before dawn
noisily but with purpose
years from now
after he and Mars
have spun away from me
this is one of the things i will remember

i will remember the winter of
my sophomore year
when i was mother
to 35 orphan lambs
i will remember
how those lambs baaahhhed
in the early morning cold
as i walked toward them with
warm bottles of milk

in the future i will remember the stars
the way the moon was
still in the sky
but mostly i will remember
the april night
when a pack of dogs
attacked my family of penco lambs
i will remember how seventeen died
and seventeen others were left
covered in their own blood
there was one lamb left untouched
born blind he had learned to follow
the other lambs by sound
i named him ciego
he had escaped
he was the only lamb that did not run

years from now
when i remember my father in the future
i will think of that lamb
not for his blindness but because he did not run
my father will
come to me in every season of my life
i will remember that he too never ran away

i will remember
in the future

how even his own family told my father
that if he chose to
he could be debt free
and not have to work so hard
in my future memory of my father
he will be told to sell his ranch
and live off of the profits
he will be told this will make him happy
in my future memory of him
he will be told to see
how the ranch
is the root of every misstep in his life

my father tells this story
of his youth and the two dreams
he had for what he would become
the first was an FBI agent
the second was a rancher
he counts himself among the luckiest of men
to have attained at least one of his dreams

my father in the future
will come to me
in the bend of the river where
it meets the meadow
in the thick and bent autumn barley
in the first trickle of ditch water in spring
his memory will surface
and persist in the rising haystacks,
the sheep spreading out along a hillside
he will rise in my memory like pale llano dust
and i too will consider him
lucky to have known and
fought for one of his two dreams

so much has gone away from antonito
so much poverty

amongst the fallen lambs
of antonito's past
my father persists despite
the FHA appraiser
projecting his crop output
as 1/3 less than his neighbor's
when all that separates the two fields
is a thin string of barbed wire
and the thick canyon
of a person's last name

i see my father rise daily
despite the same appraiser
telling him that his land is worth
less than smaller pieces
just up the road from us
he knows in his persistent heart
those ranches are white owned
he knows
in the memory of his heart
that those same ranches were
sold away by his own people
he knows there is
no happiness in their profit

in the future
i will remember my father
on the 12th day of july
he will be covered in mud
i will remember how environmentalists
destroyed the float on his stock tanks
i will remember in the future
how 30,000 gallons of water
flooded over the round lip of the trough
a mud so thick that
seven cows and seven calves

lost their lives trying
to pull themselves free of it
in the future i will see my father
fall to his muddy knees
and give a dying calf mouth to mouth

in the future
i will remember my father
drop to those same knees
in a mud and manure filled corral
to nurse a dying lamb from
its mother's swollen ubre

my father
when i remember him in the future
will have persisted through
winter storms that buried sheep alive,
through the words of "educated" books
that call him a recipient of ranch welfare

in my future memory of him
my father will be a man
i would like to become

as i remember my father in the future
Alfonzo Abeyta will rise noisily
in the stillness of every morning
the hours between darkness
and the sun's rising will belong to him
he will rise like mars on a november night
he will rise and be close to earth
he will burn in the memory
of this boy
this is where i will remember him in the future
this is where i will love him

Bob Budd is executive director of the Wyoming Wildlife and Natural Resource Trust, a statewide program that funds habitat improvement. Working in natural resource conservation can be the most rewarding, and most trying, of all endeavors. One day, at a point of particular hopelessness, a family of river otters appeared out of nowhere and changed his despair to delight. At other times it was bobcats, newborn calves—something as simple as seedlings of bluebunch wheatgrass. In this essay, he reminds us that how we choose to value such moments, and how we intend to maintain the framework in which they occur, is a challenge we must meet now, not later.

Otters Dance

Bob Budd

The worst drought in the recorded history of Wyoming began with five feet of wet snow, on the twentieth day of April 1998, when twenty cows calved somewhere in the night, and deer huddled beneath the red rocks until they were suffocated by the crushing weight of the white water. An open winter had let frost creep deep into the soil, which was as hard as the forehead of a bull. Behind the snow, a massive high-pressure system warmed the landscape, and within three days, sixty inches of snow melted and ripped anything unfrozen downhill, downstream, and gone. The red clay road never became muddy, because the snow served as insulation, then turned to water, and vanished as effectively as flushing a toilet. For the next two years, the total combined rainfall was less than what that one storm dropped on the ranch. Over the next five years, less rain fell than the average for a single year.

Those rains that came were isolated, intense, and worth little. Some covered no more than a few acres, pounding so hard that the rain raised dust, then vanished. In their wake came bizarre rainbows of dirt and water, all of which disappeared in seconds, leaving only the dusty haze. The baseball field was especially prone to deluges and lightning, and on a short June night, my oldest son, Joe, and I left a cloudburst in town, in a game he was pitching masterfully and winning, only to find ourselves bone-dry and baking in the late-day heat on the dirt road that led to the ranch. The truck was air-conditioned but our windows were open, and we watched the stream along the way, looking for beaver, muskrats, and fish feeding. It was a ritual. In the mornings, we watched the rim for bobcats, mule deer, and mountain lions. At night, we watched the bottom for creatures that emerged in the dusk.

In the grips of drought, the creek was something to cling to, a ray of hope, a reminder that this dry spell was not the end of the world, or even our own view of the world. It was also a reminder that mountains ultimately have nowhere to go but down, and rivers have little choice but to help them get there.

Joe was quiet, one of many alter egos for a man of fourteen. There were things on his mind, I knew, but few would come out now. He had great heart and tremendous strength, but he was small, and that was the only thing some people could see. Tonight, he had become large. His fastball snapped and his change-up was merciless. Even the occasional knuckleball seemed ordained to find the strike zone. He was in meticulous control, and then the rain came, the game was no decision, and he was small again. I agonized with him, for I had been in the same hard place many years before, and the memory still stung. I applauded his pitching again and told him we all emerge differently, like butterflies.

"I am not a butterfly," he growled.

"Bad example," I agreed. "I'm just trying to help."

"Holy shit!" Joe yelled.

I could have thought this was a breakthrough moment, reprimanded him for his language, or continued with the lecture, but I was following the line of his arm and staring at the creek.

"*Ho-lee* shit," I muttered.

On the far side of the creek, no more than fifty yards out, five river otters ran with an easy, incredible grace. They flowed like a watercolor masterpiece and the world became fine art, as if bronzes had suddenly awakened as acrobats. Adults in front and rear, three young in the middle, the otters undulated upstream as the creek rippled down. Pastel willows swayed in the breeze. The otters stopped, then dipped into the water, and emerged on the near side of the creek. Translucent grasses hid them momentarily, and then they burst from the streamside into the open. Water droplets flew like crystal from their oily backs, and they danced in our eyes and our minds. Within moments, the otters darted back into the creek and disappeared.

They were utterly magnificent.

"Otters!" Joe grinned. "Oh, my God, Dad. *Otters!*"

In that instant, all life ended and began again. A sullen child became an anxious young man, life rekindled in his mind, flakes of gold sparkling against the green of his eyes. The entire moment couldn't have lasted more than three minutes, but the image remains crystalline. In fact, only three things remain clear memories of that entire year. It was mercilessly hot and dry. The calves averaged more than 600 pounds apiece for the first time, and on one June night, five river otters danced up Red Canyon Creek.

For both of us the significance was visceral. Very few creatures are as stimulating as those at play, and none play with the grace and charm of otters. Their grandeur lies not in enormity but in elegance and efficiency. As they moved, their grace seemed to interpret the rhythm of the stream, to enhance the melody of water swirling through stone. The whole valley took on new meaning and excitement.

Aside from the sheer joy they provided, the otters offered something greater. For a decade my work had revolved around enhancing ecological values on the ranch, while at the same time generating an economic return. As soon as we seemed to find harmony with both, drought began to take a toll on economy, and perhaps more. Now, perhaps the premier indicator of stream system integrity, a rare and elusive top predator, was living and propagating in the heart of the ranch. Their presence signified many things: water quality, abundant fish and other food sources, adequate cover, and lots of room for otters to dance unhindered. They were validation of a team effort and cause for celebration.

Over the summer, the otters entertained many. A stranger in the grocery store showed me his grainy, distant photos of the otters at Deep Creek. One of the road crew asked me to verify what he had seen at the bridge over the river. Fishermen reported that every set of fish bones left on the bank of the stream and each track in the narrow margin of mud by the creek was cause to check it out (most belonged to beavers or raccoons, but many were otters). Suddenly, there was a bit of magic in our lives, something we could share in spite of the drought.

I wonder often what that moment was worth.

The hay we lost because of drought was worth about $25,000. The braces on Joe's teeth were worth $5,000, and his mail-order Nokona baseball glove was worth $175, most of which he saved on his own. The camera that took the digital image of the otters was worth $400. The calves were worth more than $700 each, an astronomical value for cattle. The cost to dust-guard the road was something on the order of $30,000. A gallon of diesel fuel cost $1.45. But exactly how much is a family of river otters worth?

In my brief tenure on that same stretch of dusty road, there were countless other moments. Bobcat kittens sitting on the red rim were cause for three children to be an hour late for school. A rattlesnake halted a troop of adults until ten-year-old Maggie chased it away with a stick. Two varieties of orioles, three species of grosbeaks, and a willow flycatcher were life experiences for a mother and daughter from Rock Springs. A bear weaned her cub in the corral, and the cub sat and whined at the kids when they walked to the bus in the morning. The rest of the day, he ate a lot of earthworms and whined at adults. Down at the mailbox, a savvy sage grouse hen clucked her chicks into the sagebrush while a red-tailed hawk hovered.

There was more to observation than mere pleasure. Down at the Adams place, another sage grouse hen led her chicks across the road and up a six-foot roadcut. Each of the chicks could stand in a teaspoon, but they attacked the hill with intensity, one after the other, on inch-long legs. The last of the chicks charged the hill, only to tumble backward again and again; finally, it made the summit and darted after its cackling mother.

"I think that one won't make it," came a voice from the backseat, and when I asked why not, the reply was swift and to the point.

"*Darwin* said so," said Jake.

"Darwin?" I asked.

"Yeah," was the gruff response. "Darwin. That guy *you* always talk about when dumb things die."

"Well, I think it will live," Maggie said, "because we *always* root for the underdog."

An argument ensued, one of those that must ultimately be decided by a person with some degree of wisdom, or in the absence of such, a parent. In the end, it was decreed that the chick would make it to adulthood, in spite of solid biological arguments to the contrary, because we must *always* be optimists, too.

A few years later, as observation led to science and greater understanding, we came to find that those hens and chicks had hiked fifteen miles or more to climb the hill we viewed as insurmountable, and they were only halfway on their annual journey to flight and life. One radio-tagged hen was bred in Government Draw, nested on the east side of the river, and led her brood across the river, two highways, through rural subdivisions, all the while avoiding the normal perils sage grouse encounter on a daily basis.

And again, I found myself wondering what those moments were worth. Clearly, there is no real way to find value in human emotion. A faded photo of a lost friend or worn image of a true love has tremendous value, though it may be worth little on the open market. But what value do we place on clean water, pollination, oxygen production, or predation? What are river otters really worth?

The hide of a dead river otter is a beautiful thing, sleek, shiny, and soft to the touch. A single hide is worth about $200. A family of five, then, is worth $1,000, hanging on the wall. A print of two otters, chosen for the Wyoming Conservation Stamp in 2002, is worth about $400, framed. But what are river otters worth within a landscape? What is the economic value of their presence and their role in the natural process? Pelt or painting, should they all end up hanging on walls, we stand to lose more than just otters. We *need* excited conversations and grainy photos smeared with fingerprints. We need those Holy Shit! moments.

If otters carry a greater value than the sum of their parts, the habitats they rely on are no different. We mourn the loss of crucial habitats, but we haven't captured their value any more effectively than we comprehend the worth of a living otter. We lose otters and places for them to live because our economic system has been either unwilling or unable to create effective markets for those "goods and services."

In the case of otters, and natural resources in general, there is a dichotomy between markets for needs and markets for other goods and services, including those we may not completely understand. The price of oil or gas differentiates risk and reward. Markets for cattle and hay, though volatile, allow credible estimation of profit or loss. But the market for intact aquatic systems is imperfect at best. Effective management of ranchlands may contribute mightily to water quality, soil stability, sequestration of carbon, and open space, but the market signals ranchers receive are based almost wholly on the price of livestock and hay. Lately, the only other credible economic signal has become the development price of land and water.

All economic systems are driven by cultural demands, whether highly rational, like the need for food or fuel, or highly irrational, like the lust for precious metals. In the days of Tutankhamen, Mao Tse-tung, or George Bush, markets for goods and services have always been defined by the same forces. Whether those markets are prostituted by edict or subterfuge, as in the case of oppressed societies, or poorly understood in freer societies, the demands of the people will ultimately place worth in things we value, and in doing so create a market to fill that demand.

Allowed to run unchecked or unbalanced, economic systems operate faster than natural systems can recover. China was devastated by Mao's exchange of food and human life for weapons. The American West was altered by people who arrived in a time of incredible rainfall only to find that reality was extensive drought. European nations truly valued providers of food only when they were starving. Short-term markets can emerge instantly—for pet rocks or chinchillas—and disappear just as quickly. Other markets seemingly parallel the life of the commodity.

In most cases of market development, partnerships grow out of public demand and private enterprise. Public demand led to development of massive systems for moving people, freight, and mail across the United States. Government incentives and concessions, coupled with private investment, built a rail line across the continent. Without public investment in projects of such immensity, there is no way the railroad would have been built. The same can be said for highways, which evolved from private toll roads or massive water projects, whether desirable or otherwise.

In North America, market signals on open lands have told businesses primarily to produce food and fiber, with a recent signal placing a high value on residential home sites. At the same time, cultural demands have elevated the values of open space for wildlife habitat, clean water, carbon sequestration, and other goods and services. Until recently, markets had no effective means of capitalizing and valuing conservation, primarily because the potential buyer, the public, had no means of providing capital. The parallels with development of railroads, mail service, water projects, and other large-scale endeavors are inescapable, and the precedent for public and private partnership in market development is well established.

Over time, markets mature. Today, the nation's rail freight system is privately held, efficient, and profitable. Passenger rail ultimately became a service largely funded by government, and the mail system evolved from failed private enterprise to essential government function to the present system, wherein private enterprise is rapidly forcing the traditional mail system to adapt or become obsolete. That evolution is nearly 200 years old.

As markets develop, they respond to signals. Extremes emerge, especially in early market development. Recent demand for trophy hunting in South Africa has led to release of lions and leopards from captivity to be hunted as "wild" animals. At another extreme, top-down systems in China led to wholesale harvest of timber, followed by bans on cutting of any timber. Somewhere between capitalistic extremes and radical controls lies the key to development of markets with structure and longevity, an economic radical center.

The final challenge in finding an economic center lies in understanding behavior that defies rational economic models. Simply put, ranchers and farmers behave in a manner diametrically opposed to sound economic judgment. Highly capitalized businesses with narrow returns are generally subject to liquidation. Massive gains in market share invite conversion to cash and reinvestment. Ranchers have seen costs rise, returns fall, and underlying value soar, and they still seek ways to maintain the business. People on the land put profits right back into the farm or ranch, whether by reducing debt or expanding operations. Many would be as happy to die in the corral as retire, and that creates problems as well. More than one ranching son or daughter have finally become the boss when they were well past retirement age themselves.

For years, government programs have moved away from regulatory approaches to incentive-based programs that enhance economic opportunity, while at the same time providing habitat enhancement, conservation, and development. In the private sector, habitat-based organizations continue to grow, including foundations dedicated to waterfowl, elk, mule deer, fish, wild sheep, marine life, and other species. Cattlemen and farmers have created land trusts that work in their own backyards. We are all seeking the means to balance cultural desires, economic realities, and ecological opportunities. We are striving to find a radical center.

The Sandhills of Nebraska offer a fascinating study of how some markets work in light of irrational economics. Nearly every operation started out with a 160-acre homestead. Some families survived, some did not. Some families grew while others shrank, whether from lack of births, too many deaths, or desires that led youth away from ranching. Though the amount of land and grass was finite, the need for expansion was filled by contraction. A rancher who took care of his land and paid his bills could retire on his investment after a lifetime of hard work, even without radical inflation in land prices. Usually, he could sell to either the next generation or a neighbor. Almost always, he could live out his life on the place and watch a new generation flourish.

One Sandhills rancher told me the story of part of his place, a tale of separate original homesteads. In the 1960s, his grandfather was looking to retire. The old man's daughter had married and moved away, and had no real desire or ability to return to the Sandhills, so the grandfather sold a portion of the ranch and rented the rest to a younger neighbor who needed a place to expand. Grandpa and Grandma went right on living in the house he grew up in, offering advice, picking nits, and enjoying life. The land payments and rent offered them a decent retirement.

The grandson arrived back on the ranch by chance. After his grandparents died, the young man had managed the lease, and when the neighbor was ready to retire, the most logical thing was to sell the place back to the original owner. Roles reversed, and the young man bought the homestead back along with the other place. In closing on the ranch, he found that this was the fourth or fifth time the same transaction had taken place. Moreover, he found that in nearly every direction of the compass from his home, the same patterns had prevailed for more than a century. As late as the 1990s, this system continued to work—culturally, economically, and ecologically. The reason it worked is that land values remained rooted in production of perhaps the most traditional and irrational of all commodities: grass.

At the same time, very strange things were happening to ranchers in the West. People in sleepy little places like Cortez, Colorado, or Dillon, Montana, felt as if all roads led to their backyard. One friend from southern Colorado said it happened almost overnight. Cars with ski racks, SUVs with bike racks, reggae music, brewpubs, and espresso bars popped out of the ground. Somebody had to drink all that coffee and beer. In the shadow of mountains, with the potential of trout streams, and proximity of feng shui—red rock, land, and water values skyrocketed. At the same time, many of those who liked their milk steamed and beer stout weren't overly happy with ranching. In the face of cultural and economic pressures, many ranchers took the money and ran like hell.

Those who sold feng shui needed a place to land, and they had three things left that mattered to them: family, money, and cattle. They didn't need to be close to an airport or fancy restaurant, and they damned sure didn't need a coffee barista in the morning. They needed grass and water, two things abundant in the Sandhills. They also needed a place to put the capital gains earned on their former homes, or their life's work would be taken by the Internal Revenue Service. Suddenly, the market for land in the Sandhills went to levels no one there had ever seen, and the local system came under intense pressure. Fortunately, families old and new maintained a business structure that valued large ranches, and with them ecosystem-scale natural processes.

In the land of feng shui, the price of land and water became a beacon that outshined all others, until all others were rendered useless. The "forty acres and a mule" once deemed a national ideal became a reality. This time, it came without the mule, and in many cases, without the mule deer. When the ultimate value of land becomes the view, rock and sky will define the market. But rock and sky are as inanimate as gold and diamonds. They will never, ever dance like otters along a stream.

Otters need three basic things in their lives: food, water, and shelter. Lacking one of the basics, they either explore or die. When they find all three, they are one of few species that obviously engage in play. Our world is characterized by three driving forces: ecology, economy, and culture. Lacking one of the basics, we, too, explore or die. It is not a matter of balance, really, as much as a sense of fairness, that should guide us. Finding a radical center lies in the art, or perhaps the will, to find a level of equity among all three driving forces. Historically, most human systems have tended to hold either economy or ecology somehow "most important," and always at the expense of the other. Many human systems have collapsed. Every one did so because they failed to find that elegant blend of economy and ecology that will ultimately lead to subsequent cultures.

Discovery can be empowering or frightening. Sometimes it can be both. Partial discovery is nearly always both, to the extreme. Our understanding of natural processes is no better than partial discovery, and the historical record reminds us constantly that we are still learning. Over the course of time, the landscapes of the world have changed in many ways, and they will continue to change. Natural processes demand a dynamic world—one that rises and falls to signals we cannot control, or sometimes even understand. Tectonic plates, calderas, and climate drive a planet that constantly defies constancy. On a shorter time scale, the processes of disease, fire, herbivory, drought, and flood play an equally powerful role in biological change.

We seem to comprehend that there is little we can do to truly affect geologic processes. Our understanding of shorter-term processes remains incomplete, and that seems to bother us intensely. In our core, we seemingly accept as fact the notion that meteors may strike, mountains may collapse, or a tidal wave may strike us down. At the same time, we seem to blind ourselves to agents like fire and herbivory. We defy natural processes, and sometimes we tempt fate. For every agent of change, there is a prevailing and countervailing effect. When we curtail fire, we discover the power of fire. When we implement fire, we fear the power of fire. When we do neither, we become pawns of fire.

When we fail to seek, we cease to discover. If we fail to try new things, whether from fear, complacency, or arrogance, we will cease to learn.

It is fascinating that we have mapped the great plates that move the face of the earth, found the moxie and the means to travel within the solar system, and still fallen short in our ability to understand our own organism. We've found the ability to manufacture artificial kidney function and forgotten to teach boys to write their name in the snow.

Until very recently, social sciences and history were disregarded, if not disdained, in the natural resource sciences. We spoke of the "art and science" of resource management, then threw away the art and asked the science to answer every question. In many ways, natural resource science was used to exclude humans from the ecosystem. The insight of common people was discarded as blithely as former generations ignored the wisdom of the Shoshone or the Crow.

It is a hard road, the one that calls for understanding and respect, but if we truly hope to find a radical center, we must find a way to capture the value of otters in every sense of their being. To do so, we must find relevance between their lives and our own. If the last family of otters held the key to the future of humankind, would we value the places where they lived?

I believe the answer is yes. Moreover, I believe we value those things today, fervently, but we place our faith in legislative and legal forums when we should be placing our trust in the hands of those who can ultimately make a difference. Quite honestly, otters are worth more to me than hay. If the rancher who grows the hay will continue to grow otters, he should be paid for both. He should be accorded the highest esteem for his care of the public resource, and the system should reward him mightily for his effort. We *must* find the ways and means to maintain traditional economies within natural systems or we will find ourselves lamenting what we had, instead of celebrating what we have.

We may never "find" the radical center, a concept as elusive and intoxicating as a first kiss. It is more important that we simply keep seeking that elegant balance. Maybe we need to value patter in the produce aisle more highly than pontification in public. The radical center lies in commitment to the journey more than to the outcome, to taking our lumps when they come, and to dancing with joy sometimes, like otters along a stream.

Editors and
Contributors

Aaron Abeyta is author of two poetry collections and a forthcoming novel. For his first collection, *colcha*, he received the 2001 Colorado Book Award and a 2002 American Book Award. His latest poetry collection, *as orion falls*, and his forthcoming novel, *rise, do not be afraid*, are both published by Ghost Road Press. Currently Aaron is working on a book of love poems as well as putting the finishing touches on a play and a second novel.

Julene Bair's book *One Degree West: Reflections of a Plainsdaughter* won Mid-list Press's First Series Award and Women Writing the West's Willa Award. Her essays have garnered prizes from the National Endowment for the Arts and the Wyoming Arts Council. She lives in Longmont, Colorado, where she's working on *The Whole Song*, a book about two romances: one between her Kansas farm family and the Ogallala Aquifer, and her own with another Kansas native. "Both," Julene says, "unfold like country-and-western songs. We fall in love like gangbusters, but wind up crying in the end."

Rick Bass is the author of twenty-two books. His first short story collection, *The Watch*, set in Texas, won the PEN/Nelson Algren Award, and his 2002 collection, *The Hermit's Story*, was a *Los Angeles Times* Best Book of the Year. Bass's stories have also been awarded the Pushcart Prize and the O. Henry Award and have been collected in *The Best American Short Stories*. Rick was born in Fort Worth, Texas; earned a B.S. at Utah State University; and worked as a petroleum geologist for several years. He currently lives and works in the Yaak Valley in Montana.

Bob Budd is executive director of the Wyoming Wildlife and Natural Resource Trust, a statewide program that funds habitat improvement. Previously he was director of land management and manager of Red Canyon Ranch for The Nature Conservancy in Wyoming, and executive director of the Wyoming Stock Growers' Association. He served as president of the International Society for Range Management in 2003.

Bob has worked to facilitate resource management plans with land managers in North America, Asia, and Africa, and is highly regarded as a speaker and lecturer. Bob has an M.S. in range management, a B.S. in agricultural business, and a B.S. in animal science, all from the University of Wyoming. He and his wife, Lynn, spend most of their free time following three teenagers, Joe, Jake, and Maggie, around Wyoming and the world.

Joan Chevalier is a speechwriter based in New York, where she works primarily on Wall Street. She writes speeches and editorials on international finance and politics, military strategy, U.S. politics, the environment, and America's rural communities. She hopes that her fiction, described by editors as avant garde, is published in sufficiently obscure places. When Joan began her exploration of ranch life, she was terrified of horses; now she describes her horse, Oakey, as the "great love of my life" and has begun competitive riding. For that she is most grateful to her ranching friends.

James Galvin was raised in northern Colorado. He has published several collections of poetry, most recently *Resurrection Update: Collected Poems, 1975–1997*, *Lethal Frequencies*, *Elements*, *God's Mistress*, and *Imaginary Timber*. He is also author of the critically acclaimed prose book *The Meadow* and a novel, *Fencing the Sky*. James still holds on to some land and horses near the Colorado-Wyoming line. He teaches at the Iowa Writers' Workshop.

Drum Hadley has been a rancher and a poet along the Southwest borderlands for four decades. He is cofounder of the Malpai Borderlands Group, a pioneering organization forging collaborative conservation solutions for the West. He also founded the Animas Foundation, which supports sustainable ranching in harmony with the environment. Drum is author of four books of poetry, his latest titled *Voice of the Borderlands*.

Linda Hussa and her husband, John, live on the western edge of the Great Basin, raising cattle, sheep, horses, and the hay to feed them. Linda has three poetry collections and three books of nonfiction. Her work draws from the isolated nature of ranching and her commitment to the health of rural communities. Her book *Blood Sister, I Am to These Fields* was the winner of three national awards in 2002: the Wrangler: National Cowboy and Western Heritage Museum, the Spur: Western Writers of America, and The Willa: in the name of Willa Cather by Women Writing the West.

Teresa Jordan was raised on a ranch in southeastern Wyoming and is author or editor of seven books, including the family memoir *Riding the White Horse Home* and *Cowgirls: Women of the American West*. Her work has been recognized with such awards as a National Endowment for the Arts Writing Fellowship and the Silver Pen Award from the Nevada Writers Hall of Fame. Teresa lives in Salt Lake City, Utah.

Richard L. Knight is interested in the ecological effects associated with the conversion of the Old West to the Next West. A professor of wildlife conservation at Colorado State University, he received his graduate degrees from the University of Washington and the University of Wisconsin. While at Wisconsin he was an Aldo Leopold Fellow and conducted his research at Aldo Leopold's farm, living in "The Shack." Presently he sits on a number of boards, including the Colorado Cattlemen's Agricultural Land Trust, the Quivira Coalition, the Science Board of the Malpai Borderlands Project, and The Nature Conservancy's Colorado Council. Richard was selected by the Ecological Society of America for the first cohort of Aldo Leopold Leadership Fellows, who focus on leadership in the scientific community, communicating with the media, and interacting with the business and corporate sectors. With his wife, Heather, he works with his neighbors in Livermore Valley on stewardship and community-based activities.

Page Lambert was described in *Inside/Outside Southwest* magazine as one of the most notable women writers of the contemporary West. Her memoir, *In Search of Kinship*, and novel, *Shifting Stars*, a Mountains and Plains Booksellers finalist, continue to draw high praise. A contributor to *Writing Down the River: Into the Heart of the Grand Canyon*, her River Writing Journeys for Women were featured in the January 2006 issue of *Oprah* magazine. Excerpts from Page's work appear in more than a dozen anthologies, including *Heart Shots: Women Write About Hunting, Deep West, A Literary Tour of Wyoming, Ranching West of the 100th Meridian, Leaning into the Wind,* and *The Stories That Shape Us.* After nineteen years on a small ranch in Wyoming, she now lives in the mountains above Denver.

Jeff Lee is director of the Rocky Mountain Land Library, a nonprofit educational organization whose mission is to encourage a greater awareness of the land. For the past twenty years, he has also been a bookseller at Denver's legendary Tattered Cover Book Store.

Wallace McRae has been a featured performer at the annual National Cowboy Poetry Gathering in Elko, Nevada. He has read at the National Cowboy Hall of Fame, and in 1990 became the first cowboy poet to be granted a National Heritage Award from the National Endowment for the Arts in Washington, D.C. He has published several volumes of poetry, including *Cowboy Curmudgeon and Other Poems.* Wally also finds time to manage the Rocker Six Cattle Company, his family's 30,000-acre cow and calf ranch in Forsyth, Montana.

Sharon Salisbury O'Toole lives and works on their family ranch near Savery, Wyoming. After earning a B.A. in technical journalism from Colorado State University, she and her husband turned down law school to buy old ewes. They lived for several years in remote camps. She is an editor of *The Shepherd* magazine, has published two children's books, writes editorials and articles on western issues, and writes and performs cowboy poetry. Sharon and her husband, Patrick, have three children and two grandchildren. They raise cattle, sheep, horses, and dogs in the Little Snake River Valley on the Colorado-Wyoming border.

Diane Josephy Peavey writes stories of life on her family's sheep and cattle ranch in south-central Idaho. Many of these pieces air weekly on Idaho Public Radio and are collected in her book *Bitterbrush Country: Living on the Edge of the Land.* Her writing also has appeared in numerous magazines, journals, and anthologies, including *Shadow Cat, Woven on the Wind, Crazy Woman Creek*, and *Forged in Fire*. Diane was literature director for the Idaho Commission on the Arts and cofounder of the popular Trailing of the Sheep Festival.

Laura Pritchett is author of a novel, *Sky Bridge*, and a collection of short stories, *Hell's Bottom, Colorado*, which won the Milkweed National Fiction Prize and the PEN USA Award for Fiction. Her work has also appeared in numerous magazines, including *The Sun, Orion, High Country News, 5280*, and *Pulse of the River: Colorado Writers Speak for the Endangered Cache la Poudre* (which she coedited). She holds a B.A. and an M.A. from Colorado State University and a Ph.D. in literature from Purdue University. Laura lives in northern Colorado near the small cattle ranch where she was raised, and often writes about ranchland preservation.

Nathan F. Sayre is assistant professor of geography at the University of California–Berkeley. He grew up in Iowa and first encountered ranching at Deep Springs College in eastern California; later he studied philosophy at Yale and anthropology at the University of Chicago. Twelve years of working and living in Arizona and New Mexico resulted in three books: *The New Ranch Handbook: A Guide to Restoring Western Rangelands; Ranching, Endangered Species, and Urbanization in the Southwest: Species of Capital;* and *Working Wilderness: The Malpai Borderlands Group and the Future of the Western Range.*

Mark Spragg is the author of three books. *Where Rivers Change Direction*, a memoir, won the Mountains and Plains Booksellers Award, *The Fruit of Stone*, and most recently, *An Unfinished Life* are both novels. All were top-ten BookSense selections and have been translated into fifteen languages.

Kim Stafford is founding director of the Northwest Writing Institute at Lewis and Clark College, a realm for experience in creation with words, and founder of the William Stafford Center, a program for reconciliation through language with the self, the earth, and human kin. He is author of a dozen books of poetry and prose, including *The Muses Among*

Us: Eloquent Listening and Other Pleasures of the Writer's Craft. Kim was awarded the Western States Book Award for *Having Everything Right: Essays of Place.*

Paul F. Starrs, a retired cowhand, is Regents Professor of Geography at the University of Nevada in Reno. Author of *Black Rock* (with Peter Goin) and *Let the Cowboy Ride: Cattle Ranching in the American West*, he has written more than 100 articles, chapters, and reviews and still misses cowboying, though perhaps more in thought than in deed. Paul maintains a great deal of contact with ranchers and ranch advocates, and is active in trying to find new ways for ranching and ranchers to prove useful in the urban-rural mix of the twenty-first century.

Courtney White is cofounder and executive director of the Quivira Coalition, a nonprofit organization dedicated to building bridges among ranchers, conservationists, public land managers, scientists, and others. A former archaeologist and environmental activist, he now considers himself a restorationist. His essay "The Working Wilderness: A Call for a Land Health Movement" was recently published in Wendell Berry's collection of essays entitled *The Way of Ignorance.* Courtney lives in Santa Fe, New Mexico, with his wife, two children, and a backyard full of chickens.

Paul Zarzyski left his home in Wisconsin in the 1970s to study with poet Richard Hugo at the University of Montana in Missoula. He received a master of fine arts degree and later taught creative writing. His published poetry collections include *Wolf Tracks on the Welcome Mat* and *All This Way for the Short Ride,* which received the Western Heritage Award for Poetry from the National Cowboy Hall of Fame. Paul has read at the Library of Congress, the National Cowboy Poetry Gathering in Elko, Nevada, and on Garrison Keillor's *A Prairie Home Companion.*

The Rocky Mountain Land Library's mission is to encourage a greater awareness of the land. Its 15,000-volume natural history library is especially focused on the land and communities of the Rocky Mountains. The subject range of this collection is both broad and deep, with hundreds of natural history studies of flora and fauna, and many more titles on ecology, conservation, astronomy, geology, paleontology, literature, poetry, Native American studies, and western regional history. Many titles address western land issues, while others concern the various cultures, both ancient and modern, that have inhabited the region.

The Land Library is currently engaged in a site search to provide both the shelves and proper environment for a truly unique residential land-study center for the southern Rockies. While the search continues, it is also involved in several outreach programs: the Rocky Mountain Land Series at Denver's Tattered Cover Book Store, Conversations on the Land at Colorado State University, Authors and Naturalists in the Classroom at various schools along the Front Range, the Salida Residency Program (a two-week land-study fellowship in Salida, Colorado), and a new publishing program focused on land and community in the American West.

The Rocky Mountain Land Library exists to extend everyone's knowledge of the land—waking us to the sheer miracle of life on earth while providing access to the stewardship tools we all need. The Land Library's resources and programs are designed to meet the needs of local residents as well as far-flung visitors (naturalists, researchers, writers, and artists, among others).

For more information on the Rocky Mountain Land Library, please visit its website at www.landlibrary.org.

Colorado Cattlemen's
Agricultural Land Trust

Protecting Open Space by Preserving Agriculture

Over 138 years ago, the Colorado Cattlemen's Association formed the first cattlemen's association in the nation. In 1995 they once again showcased their foresight by taking the revolutionary step of forming their own land trust. The mission of the Colorado Cattlemen's Agricultural Land Trust is to help Colorado's ranchers and farmers protect their agricultural lands, thereby continuing production for the benefit of themselves, their families, and all of Colorado's citizens.

Total developed acreage in Colorado grew from 1.3 million acres in 1970 to 2.5 million acres in 2000, a number projected to exceed 3.5 million acres by 2030. Over 80 percent of Colorado's 31-plus million acres of private land is owned by farmers and ranchers, more than half of whom are of retirement age. Ironically, the pace of development now threatens to destroy the rural qualities that attracted newcomers in the first place. As Colorado's population has boomed, CCALT have come to realize that in addition to food production, some of agriculture's greatest contributions may be in the form of wildlife habitat, scenic open spaces, cultural values, and stable communities.

CCALT's primary emphasis is to increase awareness among landowners about the use of conservation easements as a means of protecting working landscapes and as a tool for facilitating the intergenerational transfer of productive lands in a way that respects and promotes private land stewardship.

All royalties from this book will be donated to CCALT.

For more information:
Colorado Cattlemen's Agricultural Land Trust
8833 Ralston Road
Arvada, CO 80002
303-431-6422
303-421-1316
www.ccalt.org
info@ccalt.org

Credits

"The Farm" by Rick Bass, originally published in *Getting Over the Color Green: Contemporary Environmental Literature of the Southwest*, 2001, University of Airzona Press.

"A Poem from the Edge of America" by James Galvin, originally published in *God's Mistress*, 1984, HarperCollins.

"Who Are We Here, Wanting to Know?" by Drum Hadley, originally published in *Voice of the Borderlands*, 2005, Rio Nuevo Publishers.

"Things of Intrinsic Worth" by Wallace McRae, originally published in *Cowboy Curmudgeon and Other Poems*, 1992, Gibbs Smith.

"*Confronting Fear*" by Diane Josephy Peavey, originally appeared in different form in *Crazy Woman Creek*, 2004, Houghton Mifflin Company.

"*Hoof Making Contact*" by Laura Pritchett, originally published in *Colorado Review*, Summer 2004.

"Wintering" by Mark Spragg, originally published in *Where Rivers Change Direction*, 1999, University of Utah Press.

"One Sweet Evening Just This Year" by Paul Zarzyski, originally published in *Wolf Tracks on the Welcome Mat*, 2004, Carmel Publishing.